Global Supply Chain Security
and Management

Global Supply Chain Security and Management

Appraising Programs, Preventing Crimes

Darren J. Prokop
Professor of Logistics
University of Alaska Anchorage

ELSEVIER

Butterworth-Heinemann
An imprint of Elsevier
elsevier.com

Butterworth-Heinemann is an imprint of Elsevier
The Boulevard, Langford Lane, Kidlington, Oxford OX5 1GB, United Kingdom
50 Hampshire Street, 5th Floor, Cambridge, MA 02139, United States

Notices
Knowledge and best practice in this field are constantly changing. As new research and experience broaden our understanding, changes in research methods, professional practices, or medical treatment may become necessary.

Practitioners and researchers must always rely on their own experience and knowledge in evaluating and using any information, methods, compounds, or experiments described herein. In using such information or methods they should be mindful of their own safety and the safety of others, including parties for whom they have a professional responsibility.

To the fullest extent of the law, neither the Publisher nor the authors, contributors, or editors, assume any liability for any injury and/or damage to persons or property as a matter of products liability, negligence or otherwise, or from any use or operation of any methods, products, instructions, or ideas contained in the material herein.

Library of Congress Cataloging-in-Publication Data
A catalog record for this book is available from the Library of Congress

British Library Cataloguing-in-Publication Data
A catalogue record for this book is available from the British Library

ISBN: 978-0-12-800748-8

For information on all Butterworth-Heinemann publications
visit our website at https://www.elsevier.com/books-and-journals

Working together
to grow libraries in
developing countries

www.elsevier.com • www.bookaid.org

Publisher: Candice Janco
Acquisition Editor: Sara Scott
Editorial Project Manager: Emily Thomson
Production Project Manager: Paul Prasad Chandramohan
Designer: Mark Rogers

Typeset by TNQ Books and Journals

This book is dedicated to my parents, Daniel and Helen, for teaching me how to dream and to my wife, Stephanie, for fulfilling my dreams.

Contents

Preface

This book presents a detailed overview and appraisal of supply chain management in the context of security. In most books dealing with supply chain management the government is taken to be a rather nebulous entity; and when it does appear it is usually as a source of preexisting constraints to business. These constraints would involve compliance with the tax system, honoring contracts, obeying laws and regulations, etc. In other words, government is exogenous to supply chains and to the strategic planning involved in setting them up. In the context of supply chain security, however, the government is a prominent player; indeed, it has a place in the supply chain alongside traditional vendors, producers, consumers, and transportation providers. All of these players have a role in helping to secure the supply chain over various areas of vulnerability. These areas include items in storage, items on the move, and data in cyberspace.

The post-9/11 world is one punctuated by concerns over security. Furthermore, many security programs are in a state of flux. Government is playing its traditional role of policing business activity; but it is also offering a role as a supply chain partner to businesses. This is occurring not only in crime and terror prevention but in natural disaster and emergency management. This book examines government in its dual role of policing and partnering. Building government into the supply chain highlights one of the tensions that is a recurring theme in the book. Basically, partnership is an important ingredient in supply chain management; but the government has the unilateral power to decide when to replace partnership with law enforcement. How should businesses plan and implement their security procedures when dealing with this special partner?

This book highlights other tensions apart from partnering and policing. Proceeding from the establishment of trust to actual cooperation is an important journey that all supply chain partners need to make. Using a game theoretic framework the process of how to achieve a cooperative result is examined. This also includes whether or not a cooperative outcome is a stable one.

There is a tension between making decisions through algorithms and through human experience and reflection. New technologies are emerging which may revolutionize the practice of supply chain management and the gathering of information necessary to secure supply chains. Data can be shared quickly along a supply chain connected by a computer network. Items can be tracked as they move along transportation routes and through particular access points. Advances in Big Data mean that an avalanche of data can be generated over a short period of time. However, human judgment is still necessary to separate data from noise and turn the data into useful information. In other words, a technology is only as useful as the people who can use it to make meaningful decisions.

With so many moving parts to today's international supply chains and so many points where vulnerabilities may reside, it may be difficult to maintain a so-called efficient operation. Indeed, if a crisis occurs or a threat needs to be reacted to the process can be messy and chaotic. At this point it is a matter of being effective as opposed to efficient. Effectiveness involves getting a handle on the disruption/disaster as fast as possible with as much resources as possible. This is not efficient; that is, it is not a process based on market signals and negotiating. Efficiency is a luxury when normalcy has returned to the supply chain.

Finally, there is a tension between trade flows and security programs. A supply chain with no logistics (i.e., no flows of items, people, etc. between partners spread over distances) would resemble a series of isolated fortifications. Certainly, these are easier to secure; but it is the flow of trade which fuels competition, incentivizes cost control, and spurs innovation. So, if trade is good for the economy it means that securing supply chains will face the challenge of dealing with distance, multiple routes of transportation, and multiple ports of entry. Since 9/11 the US government has developed several programs to address these challenges. Many have had the benefit of private sector input. Yet, tensions remain. Security is a transaction cost; and any extra cost comes at the expense of some foregone production and trade. On the other hand, security programs are necessary to deal with any potential or realized threat. It is not an option in today's economy to leave a supply chain completely unsecured. Crime, terrorism, and natural disasters are a fact of life and resources must be devoted to dealing with them.

From a social perspective trade expands cultural awareness and may improve international political relations. On the other hand, globalized trade can lead to destabilization. An open economy is one that gives up a degree of its sovereignty. Furthermore, competition would be expanded to include foreign suppliers. Domestic suppliers, if under strain, may have to layoff domestic workers. These suppliers and workers look to their government for representation of their interests. The recent votes in the United Kingdom for exiting the European Union and in the United States for an "America First" trade policy indicate a possibility of a slowdown in the post-World War II drive to globalization (i.e., increased trade flows and multi-country supply chains). Retrenchment and trade protectionism may take root in many large economies. This could lead to widespread disruption in supply chains.

The one constant in supply chain management, however, is change. The appropriate structure of the supply chain is affected by local and global events. Supply chain security must be equally flexible. It involves finding the right mix of proactive planning and reactive actions. Not all threats can be foreseen. Decision makers face a tough challenge. This book attempts to meet this challenge though analysis and appraisal of the current system of security programs and project where improvements may be made.

CHAPTER 1

Introduction

CONTENTS

GLOBAL SUPPLY CHAIN MANAGEMENT

Definition

One can find many definitions of the term supply chain management, each with varying complexity and usually at the expense of clarity. A good way to understand the term is to think of the linkages of two or more organizations managed in such a way so as the whole is more valuable than the sum of its individual parts. The organizations to be managed are usually independent firms but they can also be departments within an organization. The linkages themselves are usually contractual relationships but they can also be the result of mergers/acquisitions. In either case

1

Global Supply Chain Security and Management. http://dx.doi.org/10.1016/B978-0-12-800748-8.00001-7

these involve negotiation and legal activities. Thus, supply chain management involves the setting up and the running of an organizational structure in order to achieve some strategic goal (e.g., profit maximization, cost minimization, increased market share, product development, etc.).

The term *global* supply chain management adds the dimension of international business into the mix. This raises issues of multinational operations, cross-cultural relationships, and a business space open to more risk (be it political or economic). A good way to understand the term is to consider the following question: What happens when an international border crosses one or more points along the supply chain? It turns out that quite a lot of things happen—even under normal circumstances. Supply chain management would be conditioned by differences in legal/contractual environments and trade barriers of various kinds. Of course, firms often find it necessary to expand their supply chain into other countries. This may be due to savings in labor cost, access to critical raw materials, or opening up new markets for goods and services. The constraints involved in cross-border trade are part of the transaction costs involved in this process. As this book discusses, securing the global supply chain and complying with government-mandated regulations are a part of these transaction costs.

Logistics Management

Supply chain management, in the simplest sense, is about organizational structure. Logistics is about *flow* within that structure. Here we mean the flow of people, material, information, money/credit, and finished goods both downstream and upstream along the supply chain in order to help facilitate the strategic goal of the supply chain.[1] Logistics management is the art and science of dealing with the constraints of time, physical space, and location. It is helpful to think of supply chain management as a skeleton and logistics management as the set of organs and veins giving vitality (via blood and oxygen flow) to the former. This is truly a symbiotic relationship—supply chain management

---

[1] In the parlance of supply chain management to move downstream is to move along the supply chain from supplier to demander. Downstream moves involve such things as: raw material moving to a production plant; subassemblies moving to storage houses; and final goods moving to distribution centers. Upstream moves involve such things as: producers giving vendors information on their supply needs; consumers sending payments to their suppliers; and consumers returning products to their suppliers for repair or because of defects.

without a good accounting of logistics is to invite problems such as bottlenecks and late deliveries. At the same time, logistics without a sound supply chain structure in which to operate is to move things while, for instance, being locked in relationships that are not cost-effective, or to move things to where they are not most needed. In other words, logistics involves tactics which can help or hurt the supply chain strategy. And each may have to adapt to the other—logistics to fulfill the supply chain's strategic goal or adjusting the strategic goal given the realities of logistics.

Time, physical space, and location. These terms introduce geography and infrastructure into the mix. This is where the rubber meets the road. The logistician and supply chain manager need to be aware of what is and what is not possible within the physical environment in which the logistics flow takes place. There are seven "rights" which logistics tries to deal with.

- Getting the right product …
- In the right quantity …
- And in the right condition …
- Delivered to the right customer …
- At the right time …
- In the right location …
- And doing all this for the right price.

Any/all of these "rights" will not be fulfilled if account is not taken for how much time it will take to move something from point A to point B—or whether point A (B) is the right origin (destination) in the first place. Moving something from point A to B is dependent on the modes of transportation that are available, the nature of the geography separating them, and the quantity and quality of the infrastructure on which it takes place.[2] Furthermore, moving across an international border is dependent on where the ports of entry are and how they are approached (i.e., which mode of transport is used).

To those who are not in the business of logistics (especially the transportation component) it is an invisible industry. In fact, we can say

---

[2] There are five modes of transportation which haul goods and materials: motor carrier, rail, air carrier, water vessel, and pipeline. Each has its own set of infrastructure and service attributes. These will filter into the cost of using the mode. These also determine what types of hauls are feasible. For a complete overview see Prokop (2014a). Because of these modal differences we can also expect different security programs and regulations which affect them.

that if the general public does not think about logistics very much it is because the logisticians are doing a good job. For example, when a customer in Anchorage, Alaska orders a book from an online retailer rarely, if ever, would he wonder about the logistics of getting the book from the Lower 48 to Alaska. The customer is simply looking forward to his book arriving by mail within a few days. He might even use a tracking number to check on the progress of the book. But rarely would the customer contemplate: where the retailer's distribution center might be; what mode of transport and carrier is to be used; what transportation hubs will the book be transshipped through and why. In other words, logistics is a black box to many. Logistics is a mysterious process by which items move from origin to destination. The public naturally prefers this process to be seamless. Of course, as discussed later, this black box has more vulnerabilities than do fixed origin and destination points; and this is why cargo theft is an important challenge to supply chain security.

International Borders and Ports of Entry

The United States has commercial trade ties with over 100 countries. From these countries imports will be arriving from thousands of airports and around 1000 ocean vessel ports. The United States may be entered legally through a port of entry such that the entrant is met by an officer of US Customs and Border Protection (CBP). This occurs at established land border crossings, water ports, and airports. The role of CBP is to regulate international commerce and to guard the border. As an open society, the US may be entered through multiple ports of entry. All global supply chains which use the US as a point of entry or exit would make use of one or more of these ports. They are vital links in these supply chains, and they are where the United States' policies on supply chain security are most often deployed.

Of the 5172 public use airports, 547 of these are known as "certificated" or those serving aircraft with seating for nine or more (see Table 1.1). These contain the set of airports most likely to be part of global supply chains into and out of the United States. Several US cities host airports which are also among the busiest in the world in terms of air cargo activity. Memphis and Louisville rank high on the list since they are the global headquarters of FedEx and UPS, respectively (see Table 1.2). Anchorage is high on the list, too, because it is a refueling point along the great circle

Table 1.1 Total Number of Airports (2011)

Total number of airports	19,782
Public use	5,172
Private use	14,339
Military	271

Source: National Transportation Statistics, 2013. Bureau of Transportation Statistics, Washington, DC, Table 1-3.

Table 1.2 Top 10 US Airports (2012) (Thousands of Metric Tons)

City	Total 2012 Cargo	World Ranking
Memphis, TN	4016	2
Anchorage, AK	2464	4
Louisville, KY	2168	7
Miami, FL	1930	11
Los Angeles, CA	1781	14
New York, NY (JFK)	1283	19
Chicago, IL (O'Hare)	1254	20
Indianapolis, IN	989	22
Newark, NJ	744	27
Atlanta, GA	646	32

Source: Top 50 Airports, 2013. Air Cargo World, p. 32.

Table 1.3 Total Number of Water Ports (2011)

Total number of water ports[a]	179
Cargo handling docks at these ports	8197

[a]Ports handling at least 250,000 short tons.
Source: Pocket Guide to Transportation, 2014. Bureau of Transportation Statistics, Washington, DC Online, Table 1-2.

route[3] for US–Asia trade. It is for this reason that much of the cargo for Anchorage is in-transit as opposed to put on or pulled off at that locale.

There are 179 water ports in the United States (see Table 1.3). The total Twenty Foot Equivalent Units (TEUs) of imports into the United States

---

[3] A great circle is the shortest distance between two points on a sphere. Air carriers fly along great circles whenever possible since there are time and/or fuel savings to be had.

Table 1.4 Top Four US Container Ports for Imports (2011) (TEUs-Thousands of Twenty-Foot Equivalent Units)[5]

City	TEUs	World Ranking
Los Angeles, CA	4059	16
Long Beach, CA	3032	23
New York–New Jersey, NY–NJ	2730	25
Georgia Ports, GA	1072	44

Source: JOC Top 25 North American Container Ports, April 29, 2013. Journal of Commerce, 46.

in 2011 was about 17 million. Thus, the top four container ports alone account for about 64% of this traffic (see Table 1.4).

In terms of land crossings into the United States, there are 117 road crossing points from Canada and 35 rail points.[4] Each Canadian province (and the Yukon territory) which is adjacent to the United States has at least one border entry point which operates 24 h per day—so that there need be no downtime in activity. Mexico shares 47 road crossing points and eight rail points. Consider the 2009 traffic data in Table 1.5. To and from Canada (just from Mexico) in 2009 imply approximately one truck every 3 (7) seconds. The data also imply 3 (21) trains per hour (day) from Canada (Mexico) in 2009. The busiest international surface trade in the world is between the United States and Canada (see Table 1.6).

Total US–Canada truck container crossings were 5,375,673 meaning that the top five accounted for 67% of this traffic. The total rail container crossings were 2,147,187 meaning that the top five accounted for 71% of this traffic. Concentrations are even stronger for US–Mexico. Total truck container crossings were 5,083,126 giving the top five truck gateways an 80% share. Total rail container crossings were 827,763 giving the top five rail gateways a 99% share (see Table 1.6).

A finished good, subassembly, raw material, conveyance, and its operator/crew will interact with CBP in some way at or before clearance at

---

[4] For a breakdown of Canada–US border crossings, see Highway Land Border Office. Canada Border Services Agency. http://www.cbsa-asfc.gc.ca/do-rb/services/hwyb-autof-eng.html. For Mexico–US border crossings, see Locate a Port of Entry. U.S. Customs and Border Protection. http://www.cbp.gov/contact/ports.

[5] A TEU is a volumetric measure of cargo activity. It is the space of a standard ocean vessel container to a depth of 20 feet. Thus, a 40-foot container is capable of transporting two TEUs.

Table 1.5 US–Canada and US–Mexico Surface Trade

US–Canada				US–Mexico			
Trucks North		Trains North		Trucks North		Trains North	
1996	2009	1994	2003	1996	2009	1994	2003
5.16 mil	4.89 mil	32,897	33,873	3.25 mil	4.29 mil	8726	7774
Trucks South		Trains South		Loaded Truck Containers		Loaded Rail Containers	
1996	2009	1996	2009	1994	2003	1994	2003
5.43 mil	5.02 mil	31,000	24,000	–	2.60 mil	–	266,469

Source: 1994/2003 data from Border Crossing Data. Bureau of Transportation Statistics, Washington, DC BTS Online. US–Canada: Tables 1a, 2a, 4a, and 5a; US–Mexico: Tables 1, 2, 4, and 5. 1996/2009 data from North American Transportation Statistics Database. Bureau of Transportation Statistics, Washington, DC BTS Online, Table 12-3.

Table 1.6 US–Canada/Mexico Border's Top Five Land Gateways: Number of Incoming Truck and Rail Container Crossings (2012)

Truck Container	2012	Rail Container	2012
US–Canada			
Detroit, MI	1,397,518	International Falls, MN	500,164
Buffalo–Niagara, NY	929,620	Port Huron, MI	397,670
Port Huron, MI	679,095	Portal, ND	238,942
Blaine, WA	327,315	Detroit, MI	218,786
Champlain-Rouse Pt., NY	281,415	Blaine, WA	165,115
US–Mexico			
Laredo, TX	1,760,041	Laredo, TX	399,839
Otay Mesa/San Ysidro, CA	781,335	Eagle Pass, TX	207,895
El Paso, TX	735,018	El Paso, TX	94,089
Hidalgo, TX	475,318	Nogales, AZ	61,395
Calexico East, CA	337,028	Brownsville, TX	54,023

Source: National Transportation Statistics, 2013. Bureau of Transportation Statistics, Washington, DC, Tables 1-52 and 1-54.

one of the ports of entry. Of course, since the point of delivery almost always lies beyond that port of entry the supply chain security issues and challenges will continue beyond that point. An import entering the United States, in terms of possible ports of entry, is dependent on the domicile of the carrier and the mode of transport the carrier uses. This point will be expanded upon below.

Global Choke Points

As commodities flow from origin to destination along sea routes, they may pass through narrow bodies of water connecting two larger bodies of water. There are about 200 straits/canals throughout the world; however, a few of these are strategic in the sense that their closure (due, say, to an act of terrorism) could create a significant disruption to world trade flows. These straits/canals would be classified as choke points. Examining a world map or globe reveals their key position. International law allows for free passage of all vessels along these routes.

- Bab el Mandeb: connects the Red Sea and the Gulf of Aden (and ultimately the Arabian Sea and Indian Ocean)
  This is a major choke point along the route of oil tanker transport from the Middle East to the oil-dependent nations of the Far East (especially, Japan).
- Bosphorus and The Dardanelles: connects the Black Sea and the Mediterranean Sea
- St. Lawrence Seaway: connects the Great Lakes and the North Atlantic Ocean
  The seaway supplanted the Erie Canal as the water-based gateway into the Great Lakes. The seaway is supported by the Welland Canal which connects Lake Ontario and Lake Erie.
- Strait of Gibraltar: connects the Mediterranean Sea and the North Atlantic Ocean
- Strait of Hormuz: connects the Persian Gulf and the Gulf of Oman (and ultimately the Arabian Sea and Indian Ocean)
  This is another major choke point along the route of oil tanker transport from the Middle East to the oil-dependent nations of the Far East (especially, Japan).
- Strait of Malacca: connects the Bay of Bengal and the South China Sea

This is the second major choke point of oil transport routes lying beyond Bab el Mandeb and the Strait of Hormuz.

- Panama Canal: connects the Atlantic Ocean and the Pacific Ocean

  The canal supplanted the Strait of Magellan just above the southern tips of Argentina and Chile and saved about 8000 nautical miles in travel. A canal widening project was completed in 2016. With larger vessels (up to 14,000 TEUs) from Asia being able to clear the locks they may possibly bypass US West Coast ports and head for the Gulf of Mexico or the US East Coast.

- Suez Canal: connects the Red Sea and Mediterranean Sea

Of course, the choke points noted above are based on physical geography. Yet, there are man-made choke points, as well. Such things as history, tradition, consumer and/or raw material proximity, presence of support infrastructure, etc. can lead trade to particular water ports, airports, or land border crossings. When these become popular, traffic congestion may set in which adds to the time cost, operational cost, and the environmental cost of using the particular ports of entry. Over time, some traffic may move to alternative ports of entry but usually not enough to mitigate the congestion problem completely.

Choke points can also be relieved when new routes are made available. Two new routing options for sea trade include the Northern Sea route along the Russia's Arctic and the Northwest Passage along Canada's Arctic. Each has its own unique set of geographic and political issues; but they have to be considered as alternatives to sub-Arctic transport among the Americas, Europe, and the Far East. Regarding air cargo, over-the-pole flights may become more prevalent in the coming years, as well. However, issues of air sovereignty (particularly Russia's) have to be taken into account.

THE IMPACT OF GLOBAL SUPPLY CHAINS

Supply chain management plays an important role in the gross domestic product (GDP) of the United States. This is why securing a steady logistical flow of imports and exports is critical. Table 1.7 gives an idea of the size

Table 1.7 Logistics and the US Economy (Billions of US Dollars)

Year	Nominal GDP	Value of Inventory	Inventory Carrying Cost	Transport Cost	Administrative Cost	Total Logistics Cost
1961	550	125 (22.7%)	31 (5.6%)	46 (8.4%)	3	80 (14.5%)
1971	1,130	236 (20.9%)	59 (5.2%)	91 (8.1%)	6	156 (13.8%)
1981	3,130	747 (23.9%)	259 (8.3%)	228 (7.3%)	19	506 (16.2%)
1991	5,990	1030 (17.2%)	256 (4.3%)	355 (5.9%)	24	635 (10.6%)
2001	10,130	1403 (13.8%)	320 (3.2%)	609 (6.0%)	37	966 (9.5%)
2014	17,458	2496 (11.8%)	476 (2.7%)	917 (5.3%)	56	1449 (8.3%)

N.B.: *Parentheses show percent of GDP.*
Source: State of Logistics Report, twelfth, thirteenth, twenty second, and twenty sixth eds. Council of Supply Chain Management Professionals, Lombard, Illinois.

of logistics activity in the United States.[6] The rise of improved inventory control technology over the years explains the fall in inventory carrying costs as a percent of GDP; and deregulation of transportation markets, along with the rise of third-party logistics providers (3PLs) and overnight air couriers explain the fall in transport costs as a percent of GDP. This is a real success story for logisticians and supply chain managers. More commodities are moving in and out of the United States than ever before and yet the total logistics costs as a share of GDP are falling.

As a result of the September 11, 2001 terrorist attacks (9/11) US airspace was shut down to commercial air traffic and land border crossings were closed. It took days before the traffic systems were up and running again. What would be the effect of a major terrorist incident on the US economy? Examining the structure of Table 1.7 reveals some immediate effects. Since the cost of doing business would go up and market uncertainty would rise, inventory carrying costs would rise since

---

[6] Since Table 1.7 relies on gross domestic product (GDP) a word of clarification is in order. GDP is the sum of the market value of all the final goods and services produced within the United States over a given period of time. Nominal GDP is not corrected for the effect of price inflation, though real GDP is. It is a measure of economic *value added* but not total economic output, also called gross output (GO). GO would include the value of raw material and intermediate goods; that is, it highlights the value of items as they are produced along supply chains. As such, GO is a better way to measure supply chain impacts. For a complete overview of GO vs. GDP see Skousen (2015).

just-in-time imports would be more costly. Transport costs would go up most likely due to rising fuel costs (especially if there were a Middle East connection to the incident). Administrative costs would rise since the compliance costs of international trade would rise. In effect, total logistics cost would rise. On top of all that, GDP would likely decline due to a recession in economic activity (especially if the incident took place at a port of entry). The total cost effect would be dependent upon the duration of any shutdown, which country or countries would be hit with trade restrictions, and to what degree transportation and cargo clearance activities would be slowed due to more regulation.

INTRODUCING THE PLAYERS

Shippers

Shippers are those who need to hire transportation services. They could be either the buyer or the seller of the commodity in question. In the context of international trade the buyer and seller become the importer and the exporter, respectively. Once it is determined who will be the shipper, this party will then contract with a transportation carrier (or an intermediary as discussed below) in order to move something from origin to destination. A minority of shippers prefer to own their own fleet of conveyances so that a contract is not necessary. While such companies enjoy flexibility of usage, they are burdened by the fixed costs of fleet management and the backhaul problem. On the other hand, supply chain security between the shipper and the "carrier" is an internal matter and the sharing of relevant information should, in theory, be easier than under external contracts. There are five modes of transport: motor carrier, rail, air carrier, water vessel, and pipeline. The mode a shipper is most likely to insource, if a private fleet is desired, is the motor carrier mode. While *Boeing* and *Airbus Industrie* (it's French) ship some of their product via airplanes, it is because they build their own. It is no accident that motor carrier is the mode most likely to comprise a private fleet. It is the mode with the lowest proportion of fixed costs; therefore, it is the most flexible given the vagaries of the production cycle and the economy.

Globalization has led to supply chains extending over international borders. Many US companies, too, are highly dependent on cross-border trade in order to keep costs down, expand into new markets, etc. These shippers are among the most prominent importers

Table 1.8 Top Six US Exporters and Importers by Water Vessel (2013) [Thousands of Twenty-Foot Equivalent Units (TEUs)]

Exporter	Industry	TEUs	Importer	Industry	TEUs
America Chung Nam	Paper/Plastics	374	Wal-Mart	Retail	732
International Paper	Paper/Packaging	173	Target	Retail	512
Denison International	Paper/Recyclables	113	Home Depot	Retail	325
DuPont	Chemicals	105	Lowe's	Retail	237
Potential Industries	Paper/Recyclables	103	Dole	Food	211
Weyerhaeuser	Forest/Wood	98	Sears	Retail	208

Source: 100 JOC Top Importers & Exporters, 2014. Journal of Commerce 15 (11), 36–54.

of goods, subassemblies and raw material. In terms of cross-border tonnage the most prominent mode of transport is water vessel—with about an 80% share of total imports and total exports (see Table 1.8).[7] Motor carrier comes in a distant second with a tonnage share of about 16% of exports and 7% of imports. Using the water vessel mode as an example, some inferences can be made about supply chain security.

The vast majority of exports here (and, indeed, among the top 100) are made up of bulk commodities. These tend to be packaged more densely than multistage manufactured items and are thus harder to hide contraband within. On the import side, by contrast, the vast majority are made up of retail items of various densities, sizes, shapes, etc. With the millions of TEUs of such containerized items entering the country each year, the supply chain security problem is quite challenging. Furthermore, the top 100 importers accounted for about 6 million TEUs in 2013 out of the total of 18.1 million.[8] In other words, about two-thirds of the TEUs of imports to the United States, by water vessel, are brought in by "small" importers. This means there are tens of thousands

---

[7] See Pocket Guide to Transportation 2009, Tables 5-5 and 5-6.
[8] Brooks (2014, p. 4).

more shipments for the government to track, and many more security plans for it to vet.

Carriers

Carriers which make their services available to shippers are known as commercial, common, or for-hire carriers. Their services are priced according to a freight rate which covers the carriage of a unit of goods (of given weight, density, configuration, and need of care) over a unit of distance over a period of time in consideration of origin and destination.

Transborder pipeline and rail use fixed-in-place infrastructure so that it is easy to know where commodities will enter the United States. Transborder pipelines are only built after successful bilateral negotiations on the part of each government concerned. Foreign-based rolling stock can cross the border only if the foreign railway had permission to extend its rail lines into the United States. Of course, the foreign rolling stock could link into a domestic line; but that requires the two railways to agree to share their infrastructure. At any rate, points of entry and exit are fixed.

Foreign-based motor carriers and water vessels are dependent on where the land and sea ports are, respectively. However, they can choose which ever port of entry and exit they would like. Such is not always the case for air carriers. This makes this mode unique. A foreign air carrier is, indeed, dependent upon where the domestic airports are; but the ports of entry and exit are regulated by the nature of what is known as a bilateral air agreement. It would spell out which airports the carrier can use and at what frequency. As of 2012, the United States had bilateral air agreements in place with 130 countries (within which are the 27 members of the European Union). Numerous countries around the world are represented among the set of bilateral partners. When the United States and its bilateral partner wish to allow *any* airport to be used as a port of entry and exit, the agreement is a policy known as Open Skies. As of 2013, the United States had 110 Open Skies agreements outstanding. Technically, though, the number is really 84 since the 27-member European Union has been negotiating with the United States as a single entity since 2008. It is interesting to note that the United States and Mexico do not have an Open Skies agreement in place while the United States does have one (since 1995) with its other North American Free Trade Agreement (NAFTA) partner, Canada. Finally, 66 out of the 84

Open Skies agreements allow for what is known as 7th Freedom of the Air on all-cargo flights.[9]

Intermediaries

Shippers and carriers represent the demand side and the supply side of the transportation market, respectively. But the market is more complicated since it is often the case that a shipper will use the services of a transportation intermediary known as a freight forwarder instead of dealing directly with a carrier. Why would this be the case? Many shippers do not have the time and/or expertise to decide such things as: which mode of transport to use; which particular carrier within that mode to contact; and how to negotiate a freight rate with the carrier. A freight forwarder performs these services on the shipper's behalf. In effect, they are a parallel to the services of a travel agent—relieving the traveler from arranging the best flights, hotel accommodations, etc.

Another benefit of these "travel agents" for freight is that they pool the shipments of many customers and can use the size of these collective shipments to possibly secure better terms and freight rates for the individual shippers. Of course, large shippers have their own negotiating power when it comes to bargaining with carriers. At any rate, from the shipper's perspective, the freight forwarder is acting like a for-hire carrier in terms of making the transportation arrangements on his behalf. However, under the law, a freight forwarder is treated like a common carrier, meaning it must have cargo insurance coverage and is required to properly handle freight loss and damage claims.

Since global supply chains run up against customs officers when their shipments enter a different country, the shippers within these chains have to make sure they are compliant with the particular country's trade rules. In addition to providing all necessary entry documents, these may involve paying trade tariffs and duties (i.e., taxes on imports), complying with non-tariff barriers (NTBs) (e.g., language, packaging, contents,

---

[9] The 7th Freedom of the Air means that a carrier operates entirely outside of its home country and transports revenue passengers/freight between foreign countries. An example of this, in a three-country setting with domicile at (A) would be: travel from country (B) to (C) to (D) or any combination of these. From a security perspective a carrier from a bilateral partner country may travel into the United States carrying cargo from a country or countries other than its own. This adds flexibility to routing but more uncertainty about cargo origin. For a complete overview of all Freedoms of the Air, see Prokop (2014b).

etc.) and rules of origin (i.e., enough value content from a trade partner to qualify for preferential treatment). This can be a daunting task, especially when multiple countries are involved in the journey of the item. Fortunately for shippers they can hire the services of an intermediary known as a customs broker. For a fee these licensed professionals deal with customs officers in order to make sure the items are cleared in a timely way and assessed for accurate tariffs and duties. It is their job to stay current with changes in international trade rules.

Government

The US federal government is the regulatory authority over international trade and interstate commerce. It sets the rules over which trade transactions are made and is the dispute settlement authority should any of the parties disagree over terms, contracts, payments, etc. In the event of crime, terrorism, or natural disaster, it is the government which is expected to coordinate and provide assistance, relief, and reconstruction.

Of course, the "federal government" is not a single entity. The executive branch may be broken down into several regulatory organizations. Their roles are to interpret the intent of laws enacted through the legislative branch and devise regulations for these purposes. The Department of Transportation (e.g., the Federal Aviation Administration and the Federal Maritime Commission) regulates the domestic and foreign carriers which operate across US borders. The Department of Agriculture and the Department of Health and Human Services (specifically, the Food and Drug Administration) regulate any foodstuffs and drugs which may enter the United States. The U.S. Coast Guard and U.S. Immigration and Customs Enforcement [both of the Department of Homeland Security (DHS)] police coastal waters and foreign workers/visitors, respectively. In fact, there are 48 federal government organizations which have jurisdiction over international trade.

In terms of supply chain security, the most prominent organization is CBP, which is a part of DHS. In fact, DHS was created post 9/11 in order to place various security entities under one roof. The best example of this was taking the customs function away from the Treasury Department and the immigration function away from the Justice Department. Also, airport security screening was taken out of the hands of private companies and, in effect, federalizing the workers who remained in the months after 9/11. Two common themes in the post-9/11 cargo security programs enacted were: (1) to gather more detailed sets of information on which to gauge risks; and (2) move the point of compliance further upstream along the supply chain and, in the case of international cargo, away from the US points of entry.

CBP has the dual role of policing the border and encouraging international commerce. In order to treat incoming shipments properly (i.e., in terms of assessing applicable tariffs, duties, etc.), CBP relies on the harmonized tariff schedule (HTS). This ten-digit code is a way to classify any incoming item in terms of its name, its use, and the materials it is made up of. While CBP has the last word on HTS codes, it is up to the shipper alone (or in consultation with its customs broker) to assign an appropriate code number. However, CBP can be asked to provide guidance if the shipper is unsure. The idea is to classify the item based on its "essential character." As a law enforcement organization, CBP requires shippers and carriers use reasonable care with handling imported items.

CBP is particularly interested in the incoming shipment's bill of lading and the cargo manifest. The bill of lading is a legally binding contract for carriage between the consignor (i.e., the shipper) and the carrier. The receiver of the shipment is known as the consignee. Carriers are required by law to prepare one before transport takes place. However, they can be written by the consignor in order to make sure handling instructions are accurately recorded. When the carrier agrees to the terms in the bill by which delivery of the shipment will take place, the consignor will agree to pay the contracted amount to the carrier for its services. The bill also serves as a document of title and receipt for the shipment (i.e., it is the consignor's proof that a carrier has taken charge of items owned, at that point, by either the buyer or the seller). The cargo manifest is a document containing information about the shipper, carrier, conveyance, points of entry/exit, itemized shipment contents, etc. It contains information used in the bill of lading but does not carry its legal weight. Both of these documents are used by CBP in order to collect data and screen shipments before arrival at ports of entry. Using a proprietary targeting algorithm to process the incoming data, CBP will determine which conveyances and shipments warrant further attention on arrival. CBP also gathers data for the 47 other federal government agencies [known as partner government agencies (PGAs)] which regulate international trade.

CBP is interested in the shipper and carrier's cargo security plans. There is a two-way approach to this. On the one hand, CBP wishes to vet the plans (within its purview as a policing organization); on the other hand, it proposes to give expedited treatment at ports of entry for shippers and carriers who meet certain criteria. This latter activity is an example of CBP wishing to act as a supply chain partner. Can these two hats be worn simultaneously? Are shippers and carriers willing to provide the level of trust necessary for effective supply chain security partnership? These are issues to be developed over subsequent chapters.

Criminals and Terrorists

Supply chain security would be relatively easy if there were no "bad guys"; unfortunately, there are. They wish their activities to go unnoticed (at least until, in the case of terrorists, they've carried out a successful mission). The criminal activity may be commercial espionage, theft, piracy, or terrorism. The reward they seek may be monetary, political, or both. They do not play by the rules of society and yet they are governed by their own special rules.

Criminal theft is relatively easy to understand. It is a matter of using ingenuity to take something from someone else because the victim is perceived to be weak or careless. Whether the criminals are motivated by contempt for the rules of society or because of sheer opportunism is beside the point; they represent a dark undercurrent to society. They rely on the structure of society to a large degree. Terrorism, however, is a different matter. It appears to be more indiscriminate and it is designed to terrorize society at large, not just the direct victim or victims. Since the overarching goal is the collapse of the society being targeted, it raises the question of existential threats in the minds of the people. Terrorists, usually very small in number relative to the society they target, try to use propaganda with their criminal actions in order to inflate the threat in the minds of society. Certainly, today's news media and the Internet are accomplices, wittingly or otherwise, in this enterprise. Successful terrorist plots receive much more coverage than those that were stopped since tragedies attract more on-air analysis, speculation, ratings, etc.

Society tends to accept that opportunistic, and even organized, crime will exist alongside the law abiding. It always has, and perhaps always will, due to the darker sides of human nature. We all wish to take advantage of opportunities, though some are governed by a sound ethics code and some are not. Terrorism is harder to abide. Yet, at the same time, the tendency to inflate the threat can lead to the wrong societal response. What is compelling to society within the post-9/11 terrorist threat is that it is global and yet is comprised of non-state actors. This gives it the feeling of being everywhere and yet nowhere in particular (apart from nebulous concentrations in the Middle East). It blends religious fervor with a willingness on the part of the terrorists to commit suicide. This gives it the feeling of something ordained and yet obscene. In short, the threat seems both existential and eschatological. But is it irrational? Well, once the other players understand the terrorists "rules" and motivations, the former are better able to counter the latter. Indeed, we would hope that government and business, when working together, will be able to counter these criminal threats.

If one believes, however, that the terrorist threat is existential it does not take too long for society to think: we have to be successful all of the time while the terrorists only have to be successful once. This is a disturbing asymmetry in favor of the bad guys, especially when many of them are willing to pay for their success with their own lives. Worse, any military or police response cannot be undertaken without restraint of law (i.e., civil laws and military codes of conduct). In other words, society resists stooping to the terrorist's level. To quote a line from the Orson Welles movie, *Touch of Evil*, the "clean" police officer's reaction to others who trample on the rules is to warn: "A policeman's job is only easy in a police state."

The criminals wish to remain on the dark side. They maintain their black markets to buy and sell, and to finance their plans. It is no surprise that they should target the black box known as logistics which, as already discussed, is a seemingly invisible activity to the general public and yet is the life blood of the supply chains which enrich the modern economy.

ISSUES AND PROBLEMS GOING FORWARD

The Games Being Played

Crime and policing are definitely not "games." But the interaction among the players is truly a *game*, at least as far as game theorists understand the word. By game theory we mean the players are making moves and each player's outcome is dependent on the moves made by the others. A player may be playing different games with each of the other players. The government is playing a game with the criminals taking the form of military and police action and reaction. But it is also playing a game with business supply chains taking the form of commercial regulation and partnership. The business supply chains are playing a game with the criminals taking the form of securing and interdicting commerce. Finally, the shippers, carriers, and intermediaries are playing commercial games with each other involving the establishment of trust so that business supply chains can become viable.

Each of the players has one important thing in common: they are all locked into a system where at least one of the other players is a source of constraints on their success. This means that given any cooperation that can be formed among the players at least one will try to blunt the positive effect of mutual cooperation. Each action can expect to be countered by a negative reaction.

Government as Both Partner and Police

The issue of trust among supply chain partners is a perennial one. With government (specifically, CBP) playing two roles, the issue becomes more complicated. As a visual aid, consider Fig. 1.1. The format of the supply chain used here follows Porter's standard model of a value chain.[10] The primary processes in Fig. 1.1 represents a simplified supply chain where value is being added from left to right. Going left to right indicates a physical flow of inputs, outputs, people, and information which culminates in a final product for a customer. Of course, going from right to left indicates a feedback process of information, capital (i.e., money and credit), and reverse logistics (i.e., returns, repairs, and material disposal). This value adding process would culminate in an enhanced profit margin for the entities within the supply chain. It may also create some other type of comparative advantage over a competing supply chain. Of course, this process is supported by the five items noted above the supply chain. What is new to the Porter model here is that government is now included as a support process; but in its capacity as a partner to the various entities along the supply chain. This partnership could take the form of expedited treatment at the border because the government has vetted and approved the security programs used along the supply chain.

Below the supply chain in Fig. 1.1 the government is again shown but here in its capacity as a source of constraints to businesses. In its role

---

[10] Porter (1985).

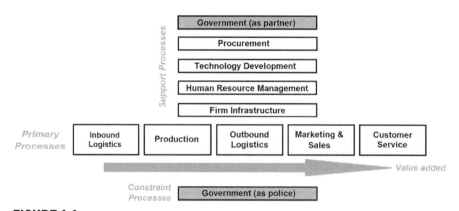

FIGURE 1.1
A supply chain with government included.

as a regulator and police it sets up transaction and compliance costs, which serve to temper the value-added process. Examples of these include compliance with HTS and NTBs.

Within the global supply chain the US border may cross through at any point—and even multiple points if inputs enter, exit, and reenter the United States as manufacturing occurs.[11]

The government, therefore, plays two roles in the global supply chain. How and when should it switch roles? How should shippers, carriers, and intermediaries deal with this? These are questions this book will deal with.

Evolving Technology

The Automated Commercial Environment is set for implementation in 2017. It is expected to be the primary way in which import and export activity will be reported. It is designed to be a "single window" of inter-action between a shipper and CBP. This means that a single submission to CBP will replace multiple submissions to other government entities.

---

[11] Consider, for example, the trade flows involved in automobile production between the United States and Canada. Sands (2009) notes that automobile parts and subassemblies cross the US–Canada border an average of seven times in order to complete a finished automobile. Global supply chains have challenged the conventional view of international trade between countries being in one or two directions at most. It is no longer a matter of raw material flowing from country A to country B and then mirrored by country B sending a finished good back to country A.

Furthermore, a paper submission will be replaced by an online submission. The intent is to streamline and speed the import clearance process, as well as the cargo screening process.

CBP, shippers, and carriers are receiving an ever-increasing flow of data—a veritable avalanche of data. It is a matter of volume and variability. In terms of volume, consider how many stock keeping units (SKUs) an importer like Wal-Mart has. As to variability, consider changes in the import flow of SKUs due to seasonal changes, structural changes in demand, or frictional changes brought about by supply chain bottlenecks or short-term changes in prices/costs. Sorting these out across all SKUs is challenging. A widely respected survey of 812 shippers and 3PLs found only about 8% and 5%, respectively, have actually developed Big Data initiatives related to supply chain management.[12] For both groups 22% are planning to implement such initiatives. For the rest of the sample, they are either familiar and apprehensive about it or simply unfamiliar with Big Data. Obviously, the concept has a lot of room to grow.

Many public and private organizations receive gigabytes of data on a daily basis and store it in databases measured in terabytes. Of course, the data means nothing unless it can be effectively analyzed. The problem is one of turning numbers into knowledge; and doing it fast enough so as to spot dangers with enough time to act on them. There is an element of risk involved in that the slowing of trade flows in order to search for illegal trade can punish legal trade. This is why cargo screening as opposed to physical inspection is the current norm.

Data analytics applied to massive databases is a fruitful area of research. It has come to the fore for a variety of reasons. First, data storage costs have dropped—giving an incentive to collect evermore. Second, evermore data can be collected quickly and, if desired, in real time. Tracking technology such as global positioning systems and radio frequency identification can pinpoint the movement of tagged objects, both large and small, along the supply chain. Shippers are developing ways to utilize this process in order to understand customer behavior down the supply chain. Yet CBP, too, wishes to use these technologies to assist it in the screening and inspection process. What the analysts do is data mine; that

---

[12] 2014 Third-Party Logistics Study: The State of Logistics Outsourcing. (2014, p. 16).
Big Data refers to data analytics or techniques used to take huge data sets and glean the necessary information from them.

is, search for patterns or relationships in a variable across time and among different variables at a point in time (i.e., time series and cross-sectional analyses). Doing this properly requires the right set of technical skills, as well as an understanding of the global supply chain and logistics activities being tracked. CBP's desire for partnering with shippers and carriers may help here; but certainly online submission of information will.

Bibliography

Border Crossing Data. Bureau of Transportation Statistics, Washington, DC BTS Online.

Brooks, C., 2014. Rising from the depths. Journal of Commerce 15 (12), 4.

Highway Land Border Office. Canada Border Services Agency, Ottawa, ON. http://www.cbsa-asfc.gc.ca/do-rb/services/hwyb-autof-eng.html.

JOC Top 25 North American Container Ports, 2013. Journal of Commerce 13 (11), 31–32.

100 JOC Top Importers & Exporters, 2014. Journal of Commerce 15 (11), 36–54.

Locate a Port of Entry. U.S. Customs and Border Protection, Washington, DC. http://www.cbp.gov/contact/ports.

National Transportation Statistics, 2013. Bureau of Transportation Statistics, Washington, DC.

North American Transportation Statistics Database. Bureau of Transportation Statistics, Washington, DC BTS Online.

Pocket Guide to Transportation, 2009. Bureau of Transportation Statistics, Washington, DC Online.

Pocket Guide to Transportation, 2014. Bureau of Transportation Statistics, Washington, DC Online.

Porter, M.E., 1985. Competitive Advantage: Creating and Sustaining Superior Performance. Simon and Schuster, New York, NY.

Prokop, D., 2014a. Transportation in business decision-making. In: Prokop, D. (Ed.), The Business of Transportation. Modes and Markets, vol. 1. Praeger, Santa Barbara, CA, pp. 1–14.

Prokop, D., 2014b. Government regulation of international air transportation. In: Peoples, J. (Ed.), The Economics of International Airline Transport (Advances in Airline Economics, Volume 4). Emerald Group Publishing Limited, Bingley, UK, pp. 45–59.

Sands, C., 2009. Toward a New Frontier: Improving the U.S.-Canadian Border. The Brookings Institution Metropolitan Policy Program, Washington, DC.

Skousen, M., 2015. The Structure of Production. New Revised Edition. New York University Press, New York, NY.

State of Logistics Report, twelfth, thirteenth, twenty second, and twenty sixth eds. Council of Supply Chain Management Professionals, Lombard, Illinois.

Third-party Logistics Study: The State of Logistics Outsourcing, 2014. Capgemini Consulting, New York, NY.

Top 50 Airports, 2013. Air Cargo World.

CHAPTER 2

The Economics of Supply Chain Security

CONTENTS

BASIC ECONOMICS FOR SUPPLY CHAIN ANALYSIS

The Market

The market may be in a physical place or it may be in cyberspace. In all cases, markets serve to bring demanders and suppliers together so that they can contemplate and complete the sales of goods and services. Physical places include any wholesale or retail outlets, stock exchanges, highway pop stands, and street vendors. Markets in cyberspace work the same way except that computer networks replace the brick-and-mortar establishments and allow for remote and often much faster trades. In

Global Supply Chain Security and Management. http://dx.doi.org/10.1016/B978-0-12-800748-8.00002-9

both cases information about product quantity, quality, and price is exchanged.

A supply chain, as an integration of two or more organizations, may also connect several markets. The common view of the market as demanders seeking out suppliers in order to transact a specific product glosses over all the submarket exchanges which led to that final product. All the raw materials and subassemblies may involve distinct market transactions between a vendor and the vendor's vendor. All the willingness to pay signals from a customer to a retailer will move upstream along the supply chain as the retailer, wholesaler, etc. each establish their own willingness to pay. Likewise, all the cost and willingness to sell signals sent from the retailer to the customer were preceded by several such signals moving downstream from the raw material supplier. In short, when it comes to supply chain management, no market can be looked at in isolation.

Then there are the logistics markets. Transportation would be the logistics activity most pertinent here. Any of the players above who do not have their own means of transportation will have to enter the market for for-hire transportation. Doing so means that price and cost signals related to the freight rate will have to be negotiated and agreed-upon. Since transportation is a transaction cost necessary for the movement of goods, these will be built into the cost and price signals among the players within the supply chain. In short, logistics markets cannot be ignored when it comes to supply chain management.

Apart from logistics are there other markets which support the supply chain? Yes. Consider third party logistics (3PLs), legal, currency, customs brokerage, etc. While transportation tends to be the most outsourced logistics activity (since most organizations do not wish to own and maintain a private fleet of conveyances) organizations along the supply chain may wish to outsource other activities. 3PLs may handle inventory control via an independent warehouse which consolidates the inventory of multiple clients. Also, marketing may be handled by specialty firms. Since market transactions with two entities, over a period of time, presupposes a contract, demanders and suppliers may need to hire legal services. Finally, if any of the supply chains cross an international border the organizations will have to deal with banks and/or international currency markets in order to deal in appropriate currencies. When the items moving along the supply chain reach a port of entry, the services of a customs broker may be needed to deal with the documentation required for customs clearance.

In summary, there are a lot of submarkets which facilitate a simple transaction between a retailer and a customer.[1] Each of these market feeds useful signals to the others running downstream and upstream along the supply chain. Coordination among these markets may be smooth or fraught with enough complexity such that information moves slowly upstream and items move slowly downstream. Well-run supply chains are noted for speed while the rest may be mired by a lack of trust, uncertainty, and inequitable sharing of risk. In all cases, however, organizations do have goals, be they narrowly focused or more holistic based on the supply chain.

Profit and Efficiency

Demanders and suppliers may have many things in common. First and foremost, both parties want to make a profit—in an economic sense. Each side wishes to receive value for money; that is, the demander wants to pay a price somewhat less than the item's value to him while the supplier wants to receive a payment which has a margin above all costs incurred in bringing the item up for sale. In this way, the transaction is win–win. When this happens the market has increased efficiency via an equitable trade.

What is meant by the term efficiency? It has two components: productive efficiency and allocative efficiency. Productive efficiency occurs when the good or service produced is done so with the least-cost combination of inputs. In this way the cost of production for a given unit of the item is as low as possible. This is a well-understood goal in any business: keep costs down; do not waste time, money, resources; etc. But the other side of the efficiency coin is allocative efficiency. Here, the idea is to produce an item that is actually desired by consumers given the costs of production. The idea can be expanded to say that the firm should organize inputs so as to produce the item which would generate the largest economic profit first. If any resources remain, and the market

---

[1] This point will be expanded in Chapter 5 with the discussion of process mapping in enterprise resource planning (ERP).

for the first product is satisfied, then expand production into the item whose profit would be second highest.

In the context of a supply chain, recall that it involves a series of integrated organizations responding to signals within their own markets. If the coordination is strong, then just as a market cannot be looked at in isolation, the goal of a given organization cannot be looked at that way either. The idea may be to maximize profit across the supply chain; and this may involve some organizations within the chain taking on higher cost than otherwise. In a simple example, if a powerful retailer squeezes its vendors for cost concessions it may end up putting so many of them out of business that the remaining vendor or vendors may have more market power through less competition. Or the striving to keep costs down might hamper vendors undertaking R&D for the eventual benefit of their retail customer. This was power projected upstream. For a downstream example, consider a vendor with a critical resource in the production process who uses this monopoly power to extract economic rents from his customers. There may be little incentive for the vendor to engage in R&D and the customers downstream bear the welfare loss of monopoly. Thus, the intent of supply chain cooperation is to have costs borne by those better able to bear it and allocate profits to those better able to utilize it. This is a challenge for any player to consider such a range of options. However, it involves trust and sharing through negotiation and contracting.

The question for supply chain security is: Can the same level of trust and sharing be engendered between the government (as either partner or police) and the players along a given supply chain? Can market incentives coexist with the nonmarket goals of government? Can government regulation of supply chain security help or hinder the marketplace? These are the overarching questions examined in subsequent chapters.

INDUSTRIAL ORGANIZATION

Competition

Consider the perfect competition model. Here competition, in an economic sense, gives incentives for firms to control their costs. Market share cannot be taken for granted in competitive markets. The prices of homogenous products such as wheat, crude oil, gold, etc. are set in world markets. As such, firms usually compete only in terms of keeping costs low enough to insure an accounting profit as opposed to an economic profit.

Now consider monopolistic competition. Here competition would also take place in terms of product differentiation. In this way, firms could enjoy economic profits as long as they attract a devoted section of the customer base. The emphasis here is on marketing and differential quality of a firm's product.

Finally, consider oligopolies. In this case firms would form alliances so as to control the market. The Organization of Petroleum Exporting Countries (OPEC) is a prominent example of a cartel's attempt to control the world price of crude oil. The success of OPEC is contingent on the cartel members' discipline (i.e., to restrict supply to agreed-to levels). Since the product is homogenous, and naturally suited to perfect competition, incentives to cheat exist. While cartels are illegal in the United States, sometimes oligopolies can be more subtle. For example, the dividing up of commercial airlines into regional hubs allows the carriers to try to generate economic profits by each tacitly agreeing to limit competition in the other's hub.

Supply Chain Competition
It is more accurate to say it is not firms that compete; rather, their supply chains compete. This means that Wal-Mart's supply chain competes with Target's, General Motor's competes with Toyota's, etc. Competition may be in the form of lower total logistics costs across supply chain partners. This could be achieved through better negotiation and contracting among the supply chain partners. It could also be achieved through superior logistics; that is, speed to market. Economies of scale, if achieved, could also control costs and, furthermore, increase market share. It could also be achieved via a bulk discount on shipping costs. Finally, competition through R&D could lead to a "better" product within one chain versus the competitors'. In this case, "better" could mean the product occupies a niche which is preferred by the consumer base.

GAME THEORY
Players
A "game," in the game theory sense, is any interaction between two or more players with each trying to achieve a particular outcome. The

players may be partners or adversaries. They may make their "moves" simultaneously (e.g., rock-paper-scissors; placing sealed bids at an auction; or submitting a business plan in competition for a contract) or in sequence (e.g., chess; placing bids in an English auction; or negotiating over a price).

The outcome of the game may have a distinct winner and loser—known as a zero-sum game in that one's gain is matched by another's loss. On the other hand, a game could be win–win, or positive sum. This occurs when a player's actions (even when performed out of pure self-interest) have spill-over benefits for other players. Finally, a game could be lose–lose, or negative sum. This occurs when each side is spiteful/vengeful to the other to the point where self-interest has morphed into each side trying to outdo the other in terms of damage. An example here is nuclear war—where no side "wins" and most/all bystanders are negatively affected. Of course, in today's world of religious-based terrorism it is possible that some terrorist may disagree with that view since the "win" was defined in earthly and not eschatological terms. However, this raises important assumptions in game theory. These are: (1) each side is able to understand the other; and (2) each side acts rationally (i.e., each seeks to maximize his payoffs). Understanding does not imply agreement, empathy, or sympathy of one for another; it means an awareness of motivations. Payoffs may be monetary or they may be utils of satisfaction. With all of this in place, strategic calculations can be made.

Competition to Cooperation

Game theory lends itself quite well to demonstrating when, if at all, a competitive situation may morph into a cooperative one. In this case it is not referring to competition in the marketplace but rather to competition between interacting players who might gain from cooperating. Why do they not simply cooperate? The reason may be lack of trust or simple selfishness (i.e., trying to gain more from not cooperating as opposed to gaining less from cooperating).

Interaction among players is important. It is also important that this interaction be ongoing or, at least, with a reasonable expectation of future interaction. It is not so easy to snub someone if you expect to see him again. All these factors may allow seemingly competitive situations to evolve into cooperative ones as discussed ahead. Trust can be learned; and it may be engendered through rewards or through punishments.

Information

They say that knowledge is power; and one cannot have knowledge without first having information. But they also say that ignorance is bliss; and no news is good news. What should you believe? All this shows is that there is a cliché for everything. But, as it turns out, information (or the lack thereof) is an important factor in game theory. The path from data to wisdom is discussed in Chapter 5 and information and knowledge are intermediate steps along the way. Information is the first item along this chain of knowledge which is considered meaningful to the decision maker. It is meaningful because all the noise (i.e., meaningless data) has been separated from the dataset which can be processed into information. The ability to isolate and ignore noise is an important task but not an easy one. When more data comes in, there is a tendency to give it scrutiny, possibly weigh some of it against the rest, etc. This means there are more possibilities to incorporate noise, mistaking it for information rather than simply ignoring it. The players of these games are struggling with two distinct forces:

- Top-down effects (which are deductive; based on theory, instincts, and preconceptions)
- Bottom-up effects (which are inductive, based on experience)

Top-down effects are at their strongest when bottom-up effects are at their weakest. This means that when a lot of noise is present (or, in an audio-visual sense, the images were blurred or the sound was droning and constant) the decision makers will hold to their preconceptions longer than otherwise. In other words, they will try to make sense out of the noise, or try to attach determination to otherwise random events. One way to cut down the noise is to look at the data in smaller doses rather than in a constant stream. This gives the mind a chance to pause and ponder.[2] Standing back allows for reflection and a chance to digest the incoming material.

---

[2] Brewer and Loschky (2005, p. 36) note a prominent experiment where viewers who received more vague stimuli than others were less likely to correctly identify an object compared to the other viewers. In other words, when both groups were given a blurry picture of a fire hydrant, those who saw the picture come into focus over more steps could not identify it at the same point of clarity as could the other group. The implication is that more exposure to blurriness (i.e., noise) overloaded their bottom-up processing compared to the other group. Brewer and Loschky discuss this top-down and bottom-up interplay at length.

If someone is driving from point A to B along a highway, it is easy to ignore the noise (e.g., shrubs growing along the side of the road, the shapes of clouds in the sky, and the cars on the other side of the concrete median going in the opposite direction) and concentrate on the meaningful data (e.g., road conditions, speed, and signs) because it is a controlled environment governed by well-known rules. The environment of supply chain security does not have all of the players willing to follow the rules because there are some players who are truly adversarial. Using the driving metaphor it is like placing a stop sign in the middle of a block of houses rather than at a road intersection. The top-down effects are stronger than the bottom-up ones and such stop signs are easy to miss. In other words, the drivers need to prepare for outlying events.[3] Stimuli must be clear and strong in order to dislodge lazy top-down effects.

It would be nice if all the players had equal and full information about each other; but it would be unlikely. The real world is characterized by asymmetric information (where each side has different packets of information that others are not aware of). Such a situation is prone to miscalculations and mistakes. Yet, in certain cases, it can actually enhance stability. For example, Schelling (1966) showed that in nuclear stand-offs the fear of retaliation outweighed the perceived ability to withstand retaliation. With this in mind, one player might prefer to keep his enemy guessing about his intentions rather than reveal his hand completely. This was very much the case of the twentieth century's Cold War where retaliation (i.e., mutual assured destruction, MAD) acted as a check to unrestrained adventurism.

Today's situation is characterized by so-called asymmetric warfare of non-state actors threatening well-defined nation states. What is also different is that the war is not "cold," the enemy is not monolithic, and its locale is less defined. It appears that the information asymmetry is even more pronounced; therefore, the situation is apparently worse for the United States and its economy. Of course, such thinking is based more on emotion and the uncertainly involved in trying to understand a new enemy instead of a familiar one. Furthermore, non-state actors are not an existential threat (though they could cause serious damage to the economy and parts of the social fabric).

---

[3] Brewer and Loschky (2005, pp. 41–42).

Consider other classifications of information in game theory. Perfect and imperfect information refer to whether or not one player is aware of every move the other player made. In sequential games it is fairly easy to imagine each player seeing a move and then reacting to it. In simultaneous games, on the other hand, a player can only move so as to achieve a desired result in tandem with what move the other player might make. Of course, this other player is making the same calculation. The collective outcome may or may not be to each player's liking as will be seen ahead.

While perfect versus imperfect information concerns knowledge of the moves the players made, complete versus incomplete information concerns knowledge about the structure of the game being played. Here we are determining whether or not one player makes a move knowing the payoffs the other player faces and, given those payoffs, how the other player would react to the move. Simply put, a player has perfect information if he sees everything that it going on; and he has complete information if he knows all of his options and those of the other players (i.e., he knows the rules of the game—meaning the environment and the state of nature). In the real world, most games are going to be incomplete. Furthermore, since each player is aware of his own unique abilities and payoffs to a better degree than others, this incomplete information is usually asymmetrically distributed.

Of course, there is no reason to assume that perfect and complete information coexist. Why? All we need is to introduce an element of chance and suddenly, while we may see every move the other players make, we do not know exactly why they did what they did. From our perspective the game's information set is perfect but incomplete. This difference can also happen when we consider non-state criminal actors. We can feel the effects of their moves; we can even anticipate their moves; but we may not know where they came from (especially in cyberspace). We may not know the initial move of the game in terms of the non-state actor's locale. We may not know where they will relocate to afterward. All we know is the possible set of locales. Thus, the information is imperfect but complete.

The non-state criminal actors use terrorism and want to be given credit for it. They want their moves seen by the public. Since 9/11, they have continued to target commercial airlines and areas of tourism and commerce. Their moves have a degree of certainty: mass casualties at an economic target. For all the fluidity of non-state actors we know reasonably well that the threats originate in the Middle East as does much of the lone wolf cyber-activity aimed at the United States. The United States, too, attacks such that we know it was a military strike emerging from a fixed base.

So, it appears that this game would be one of perfect and complete information. However, the term asymmetric warfare is accurate here. Civilian lone wolves are acting behind enemy lines probing for vulnerabilities. The United States responds to this threat with policing and intelligence. The non-state actors in the Middle East act in a quasi-military fashion and cyber-spread propaganda around the world. The US military responses are confined to overseas threats with both military formations and special operations forces. So, in the worst case, moves are hidden and we do not know what "rules" trigger a strike from one side to the other. Such a game would be imperfect, incomplete, and with asymmetric information.

The challenge for each side is to gather intelligence of a quality to change that situation. This means: (1) learn every move the other makes; and (2) learn what rules they apply to point (1). The question is how do we play games with these information variations and what do the outcomes look like?

Credible Commitment

One way to make a commitment credible (whether in the form of a promise or a threat to do something) is to set it within a contract between the two parties with the expectation that the state will enforce the contract. In multistage games, contracts are one way to solve what is known as the credibility problem. But what if the state itself is a party in the game? While it is possible for a private entity to establish a contract with the government there are some things that the latter may not wish to enshrine in a contract. For example, the Department of Homeland Security (in particular, Customs and Border Protection) wishes the flexibility to decide when to switch roles from supply chain partner to police enforcer. A less potent way around a refusal to contract is for the government to stake its reputation (in this case, a publicly stated policy for secure supply chains) as an incentive for shippers and carriers to join their voluntary programs and devote the time and expense to comply with other trade regulations. An example of such a program is Customs-Trade Partnership Against Terrorism (C-TPAT) which is discussed in

Shipper/Carrier

		Partner	Independent
Government	Partner	(20,20) (r,r)	(0,30) (s,t)
	Police	(30,0) (t,s)	(10,10) (p,p)

Payoff Order: (Government, Shipper/Carrier)
Where: t>r>p>s and r>(t+s)/2

FIGURE 2.1
Security programs as a prisoners' dilemma game.

Chapter 4. The government has a choice between being a partner or a police officer while the shippers/carriers choose between being a partner or independent.[4] Partnership means mutual cooperation; but each party in a one-shot game has an incentive to deviate from cooperation in order to ensure his own narrow goals. This is a classic problem of the prisoners' dilemma. In a one-shot game, the Nash equilibrium is not Pareto-optimal as shown in Fig. 2.1.[5]

The numbers shown in Fig. 2.1 are only important in an ordinal sense. In fact, any set of payoffs which follows the inequalities shown in the figure will constitute a prisoners' dilemma. The Nash equilibrium for this game is for the government to police the shippers/carriers and the

---

[4] The discussion of C-TPAT in Chapter 4 will show that shipper and carrier partnership with the government is not without cost. Independence avoids these program costs but does take on the risk of more scrutiny at ports of entry.

[5] The Nash equilibrium is defined as a situation where there is no gain for a player to deviate from a position provided that no other player deviates. The way to find the Nash equilibrium is for a player to choose the best move given another player's perceived move. For example, consider the government's choices in this prisoners' dilemma game. If the shipper/carrier chooses to "partner" then "police" would be the best move since the payoff of 30 is greater than 20. But the choice of "police" would also be the best move if the shipper/carrier chose to be "independent" since 10 is greater than zero. Similarly, the shipper/carrier would choose to be "independent" no matter what choice the government is perceived to make. So, the players settle for (police, independent) or (10, 10) and this becomes the Nash equilibrium. Pareto-optimality, on the other hand, defines a situation where one player cannot be made better off without hurting another. Since (partner, partner) is a combination where both sides gain (since payoffs of 20 are greater than 10), it is called Pareto-improving. But once Pareto-optimality is achieved there is no other choice a player can make which is Pareto-improving. The problem is that any player, alone, has an incentive to deviate from this Pareto-optimal position in this simple one-shot game; and this is why (partner, partner) is not a Nash equilibrium.

latter to remain independent of any option to partner. However, partnership would improve the situation collectively. A way to achieve that Pareto-improving result is through continued play where each player uses a tit-for-tat strategy of partnering in a game but punishing in the next game if partnership were not reciprocated. One side telling the other of this commitment to reap the reward of cooperation (r) is credible because it can be shown that the stream of payoffs from this strategy exceeds one of mutual punishment (p) or switching between the temptation to not cooperate (t) and the sorrow of unrequited partnership (s).[6] It is also important that both players consider the game to be open-ended (i.e., infinite); otherwise, the incentive to cooperate diminishes through backward induction from the final game.

Multiple Equilibria and the Need to Coordinate

The prisoners' dilemma game, like many games with only one Nash equilibrium, is popular with game theorists because it is very deterministic. However, when a game has two or more Nash equilibria a different approach must be used. Thomas Schelling's advice was to allow for environmental and cultural factors to influence players' decision making in order to create what he called a focal point.[7] Focal points exist when the players share a culture or particular experience and would settle on the same or similar rules of thumb to make decisions.

In this way, it is possible to depart from the prisoners' dilemma model characterized by self-interest and build in, for example, the concept of a safe society. Shippers and carriers may be willing to accept the cost of regulations and compliance, because they see them as helping make not only more secure supply chains but a safer society as a result. The government, as well, may feel that partnership is preferable to policing. Fig. 2.2 outlines this coordination game.

The difference here is this game has two Nash equilibria: (partner, partner) and (police, independent). However (partner, partner) is a Pareto-optimal choice. If 9/11 and the shared experience of border,

---

[6] The establishment of trust and partnership is fundamental in effective supply chain management. In game theory trust and cooperation can be seen as an evolutionary process through multiple encounters of players. For a complete analysis see Axelrod (1984).
[7] See Schelling (1960).

	Shipper/Carrier	
	Partner	Independent
Government Partner	(30,30) (r,r)	(0,20) (s,t)
Police	(20,0) (t,s)	(10,10) (p,p)

Payoff Order: (Government, Shipper/Carrier)
Where: r>t>p>s

FIGURE 2.2
Security programs as a coordination game.

port, and airport closings truly changed the culture and environment on the part of the players it should be easier to reach (partner, partner) even in a one-shot game. This collective way of thinking acts as a focal point when the player chooses what to do. Of course, it is important to point out that while (police, independent) is an inferior equilibrium there is no reason to rule it out. All it takes is for one side to mistrust the intentions of the other.[8] For example, the shipper/carrier may think that the government thinks it is not following through sufficiently on partnership. The shipper/carrier would then expect increased policing to be the response; and this could lead the shipper/carrier to, indeed, become independent. This leads to (police, independent). Such a negative feedback loop can be broken in two ways: (1) an authority figure rises to break the discord; or (2) there is a change in the environment or state of nature. In a democracy, point (1) really means the government, since this is an economy-wide supply chain security game being played. However, the government is an actual player in this game. It is not an outside party trying to mediate a dispute. Point (2) is more plausible; but it is an exogenous factor. In this sense, the supply chain security game has been one that reacts to changes in the state of nature.

Do focal points apply even to the criminals and terrorists? Yes, to the extent they are rational and their own cultural norms can be determined. Chapter 3 looks at the nature of the threats and Chapter 5

---

[8] Myerson (2009, p. 1113) calls this a "social pathology." Only social change can improve such negative feedback loops.

discusses the new frontier of cybercrime. Starting with the premise that criminals and terrorists look for vulnerabilities and then strike when they see an opportunity to do so, it becomes a matter of the government and the private sector identifying likely targets and the times and places when both vulnerability and opportunity seem to be at their highest. While the government and the private sector are trying to deter criminals, they are, in a sense, playing a coordination game with them as well. They need to make the criminals and terrorists plainly aware of their programs and that they are designed to limit vulnerability and opportunity. Any weak or mixed signals do not enhance deterrence, and strong signals require coordination so that their adversaries receive them.

The State of Nature in United States' Security

What can be said about the environment (both physically and socially) in which the United States resides? The United States is a unique and fortunate country. Uniqueness comes in the form of its government (i.e., constitutionally divided power which incentivizes deliberation and either compromise or deadlock), dedication to individual freedom, and super power status. Fortune comes in the form of the inventiveness of its citizens, flexibility of its economy, and its geopolitics. Specifically, the two vast oceans separating North America from Europe and Asia have given the United States a sense of natural protection. Of course, this sense can lead to complacency until it is shaken by events which are foreign in origin (in particular, the attack on Pearl Harbor on December 7, 1941 and the terrorist attacks on September 11, 2001).

Government and the general public can fall into the mind-set that if something is unfamiliar or unthinkable (i.e., a sneak attack by a hostile power or a terrorist event using airplanes as weapons of mass destruction instead of the more familiar hijack tactic) it must be improbable. In other words, the absence of proof is taken to be proof of the absence.[9] However, if one thinks like one's adversary or enemy, then these events should not be too surprising. For example, the Japanese "sneak attack" was a reaction to the US oil embargo on the Japanese economy due to its imperialist actions in Far East Asia. The al Qaeda

---

[9] Taleb (2007, p. 310) calls this the round-trip fallacy meaning confusing the absence of evidence to mean there is evidence of absence.

terrorist network had earlier made a quasi-declaration of war on the United States, and sympathetic elements (who trained in al Qaeda camps in Afghanistan) had launched an earlier bomb attack on the World Trade Center in New York in 1993.[10] Of course, this is all hindsight and leaves aside the many other leads which were followed by intelligence officials that led nowhere. But the elements were, indeed, there. The challenge is to identify the unfamiliar or unthinkable and actually begin to think about it and build it into decision-making processes. One may not develop precise answers right away, but it is better than remaining complacent.[11]

Statistically, these events can be considered to be "black swans" or outliers which do not seem to fit within a normal (Gaussian) distribution of probable outcomes. This means they tend to be ignored by decision makers and the general public.[12] These types of events have three properties: (1) they were not expected when they occurred; (2) their impacts (whether positive or negative) were very high (i.e., game-changing); and (3) in hindsight they seemed all too predictable. In other words, they are game-changers because they have changed the state of nature. For example, the Pearl Harbor attack set the United States, due to its entry into World War II, on its path to super power status and continued global engagement. The 9/11 attacks changed the aviation industry, the procedures of international trade, and opened up an ongoing national debate on security surveillance by the government. In other words, a country with a long history of open trade has increased its desire for secure trade. Of course, the reaction to a black swan could be either overreactive or underreactive (meaning, in both cases, that policy makers and society may not learn the correct lessons from them). For

---

[10] For a complete history of the events leading up to 9/11, see The 9/11 Commission Report (2004).

[11] In effect, to use the now famous terms attributed to former Secretary of Defense Donald Rumsfeld, it is the process of turning an "unknown unknown" into a "known unknown." In other words, one recognizes that there are some things we do not know very well but ought to. Furthermore, once it becomes understood, if indeed it can be, then this "known unknown" becomes a "known known."

[12] However, data on terrorist events and natural disasters such as earthquakes do show a functional relationship known as the power-law distribution when the number of events is plotted against their magnitude. Such magnitudes include the number of casualties in terrorist events or the size of the earthquake on the Richter scale. See Silver (2012, pp. 428–437).

example, was the regulatory reaction to the financial crisis of 2008 the correct one or are the seeds for the next crisis growing despite these new regulations? The same question can be asked about the post-9/11 regulations on trade and transportation. Is the success of a human-caused black swan event a deliberate outcome or should it be considered random? Were the actions taken to prevent another 9/11, given that no such event has occurred in over 15 years, evidence of success or simply a *post hoc, ergo propter hoc* fallacy? This idea of causation will be looked at more closely in Chapters 4 and 5.

ISSUES AND PROBLEMS GOING FORWARD

Many of the threats to supply chains are indeed true black swan events (i.e., natural disasters which cannot be predicted at all until, perhaps, they are imminent) while some are human caused and may be predicted after a lot of careful thought. Supply chain threats are discussed in Chapter 3. Technology is being used in an attempt to come to the rescue. Big Data and the connecting of business and government through enterprise resource planning (ERP) are meant to increase the data available and share it seamlessly. Will they work? This is the topic of Chapter 5. The federal government is requiring more trade and security regulations but attempting to make compliance easier through a paperless electronic environment. The status of these programs is covered in Chapter 4. The question is: Do these programs lead themselves to a prisoners' dilemma game or a coordination game? Chapter 6 explores how business and government are trying to coordinate within the regulatory process. It is worthwhile to keep the games discussed in this chapter in mind when considering the issues these different parties are wrestling with.

Bibliography

Axelrod, R., 1984. The Evolution of Cooperation. Basic Books, New York, NY.

Brewer, W.F., Loschky, L., 2005. Top-down and bottom-up influences on observation: evidence from cognitive psychology and the history of science. In: Raftopoulos, A. (Ed.), Cognitive Penetrability of Perception: Attention, Action, Strategies, and Bottom-Up Constraints. Nova Science Publishers, Inc., New York, NY, pp. 31–47.

Myerson, R.B., 2009. Learning from Schelling's strategy of conflict. Journal of Economic Literature 47 (4), 1109–1125.

The 9/11 Commission Report. , 2004. W.W. Norton and Company, Inc., New York, NY. https://9-11commission.gov/report/.

Schelling, T.C., 1960. The Strategy of Conflict. Harvard University Press, Cambridge, MA.

Schelling, T.C., 1966. Arms and Influence. Yale University Press, New Haven, CT.

Silver, N., 2012. The Signal and the Noise: Why So Many Predictions Fail—But Some Don't. The Penguin Press, New York, NY.

Taleb, N.N., 2007. The Black Swan: The Impact of the Highly Improbable. Random House, New York, NY.

Threats to Supply Chains

HUMAN THREATS

Theft at Storage and in Transit

Supply chains are all about adding value along a production process. Something of value is always a target for theft. The raw materials, subassemblies, and finished goods which make up the outputs of a supply chain must be procured, stored, and distributed. Items inbound from a vendor, in storage in a warehouse, or outbound to a customer are the points along the supply chain links where thefts might occur. Where to build in the security is a challenge. It is said that "cargo at rest is cargo at risk." Of course, if one always keeps cargo on the move it enters the "black box" known as transportation. It is easier to secure a fixed position than it is a series of positions along a journey. This represents an unpleasant trade-off.

Global Supply Chain Security and Management. http://dx.doi.org/10.1016/B978-0-12-800748-8.00003-0

Storage is typically in a fixed place within some facility (e.g., a factory, warehouse, or retail outlet) and the degree of fortification can run the range from a simple "no trespassing" sign all the way to a quasi-military installation. It is relatively easy to defend items in storage because guards can remain in one place, the perimeter can be fortified, and surveillance cameras and motion detectors can be used to scan the surroundings.[1] The criminals, on the other hand, have to approach without detection, breach security, and then escape with the items in tow. Another option is to bribe employees on the inside. They could place stolen items in dumpsters during the day and these could be retrieved by the thieves at night. Another option is to place the stolen items in conveyances alongside legitimate shipments. The stolen items could be retrieved when the shipment is delivered to another warehouse.

Inbound and outbound items have one thing in common: they are in transport. Compared to storage, the tables have now turned. The criminals may lie in wait and spring their trap on a relatively more vulnerable conveyance passing by in a more open space. Or they may follow a shipment in transit and wait for an opportune time to act. The maxim "cargo at rest is cargo at risk" certainly applies. For these reasons cargo theft is of major concern. Note that piracy is a special case which will be discussed below. The FBI estimates that cargo theft in the United States adds up to losses of around $30 billion annually with consumer prices for the items rising up to 20% more as a result.[2] Surprisingly, the FBI has only been in the cargo-theft data collecting business since 2005, with the passage of the USA Patriot Improvement and Reauthorization Act. It aggregates theft data collected at federal, state, and local levels.[3] The number of cargo thefts reported in 2012 was 946. The FBI's Uniform Crime Reporting Program offers the following definition of cargo theft: "The criminal taking of any cargo including, but not limited to, goods, chattels, money, or baggage that constitutes in whole or in part, a commercial shipment of freight moving in commerce … (and) for purposes of this definition, cargo shall be deemed as 'moving in commerce' at all

[1] For an overview of perimeter security, and the issues surrounding fences/walls, windows/doors, locks/keys, biometrics, surveillance devices, and alarms see Fischer et al. (2013), Chapters 8–10.
[2] Federal Bureau of Investigation (2010).
[3] Federal Bureau of Investigation (2015).

points between the point of origin and the final destination, regardless of any temporary stop while awaiting transshipment [*sic*]."

An annual cargo theft of $30 billion in the United States is a lot of value stolen. McNicholas (2008) provides a lower estimate of $10 billion for 2000 while Hoffer (2010) comes up with a worldwide figure of 1% of the gross domestic product (GDP) for all cargo and supply chain theft.[4] If we take Hoffer's estimate and combine it with the World Bank's estimate of world GDP at $78 trillion (2014) the total amount stolen across the world is $780 billion. This, too, looks like a lot of value taken out of the legitimate economy and placed into the black market. Of course, it is important to note that such figures are rough estimates. They involve data based on reports to law enforcement. Of course, not all theft is reported; therefore, aggregates may be misleading. Another approach is to take a sample and report averages. Some security companies publish results gathered from their industry clients. For example, *LoJack SCI* publishes quarterly cargo theft activity based on their "trusted community" of manufacturers, shippers, distributors, consignees, and law enforcement and insurance entities. Their report for the first quarter of 2015 identified an average loss per incident of about $196,000 based on 84 reported incidents across the United States. Sixty percent of these losses took place between Friday–Monday.[5]

What is being stolen? Since such crimes tend to be ones of opportunity, it makes sense to target items with high value-to-weight ratios (to improve reward relative to risk), are not too fragile or bulky (lest they impede the get-away), and have established black markets (in order to speed the resale of the stolen goods). Alcohol, tobacco, pharmaceuticals, and consumer electronics are typical examples. Hoffer (2010) proposes a "product risk factor" to help shippers and carriers assess what can increase risk. The five factors are:

- Timing of the product shipment
- Product availability in the marketplace
- Level of product protection when shipment is unattended
- Product demand in the marketplace
- Profit from the sale of the product

---

[4] McNicholas (2008, p. 145) and Hoffer (2010, p. 19).
[5] See LoJack SCI (2015). Other companies which collect private data are *FreightWatch International* and *CargoNet*.

Using these five components, one can imagine the riskiest shipments to be those with: very predictable shipping times and routes, are in short supply but in high demand, and are very profitable but are not under constant surveillance when being transported.[6] The higher the weighted average value of product factor risk, the more imperative a supply chain security plan is for the product and its transport.

Criminals compared to terrorists, we recall from Chapter 1, work within society and rely on its rules in order to circumvent them for profit. Their black market or underground economy works in parallel with the legitimate, above board market economy. Consider how organized crime groups can steal cargo in a quite business-like way. Once the conveyance (say, a truck or tractor-trailer) and/or the container(s) holding the shipment is stolen, it is just a matter of turning the stolen items into seemingly legitimate items. With a little advance preparation, the criminals should already have set up a legitimate company as a front. It could be a shell merely existing on paper to give a name to the company when it deals with legitimate companies downstream along the supply chain. Of course, the front could be in the form of a parallel legitimate company (e.g., a criminal ring channeling stolen automobile parts through its own manufacturing business). This may be convenient because if the parts are sold to legitimate businesses then records would be kept. It is relatively easy to mask the sale of stolen items when placed among those duly manufactured by the company. If instead the criminal ring did not maintain a productive business and simply wanted to sell the items, it may wish to sell for cash only and not issue receipts. If these schemes lead to large inflows of cash, the criminal ring may, nonetheless, feel the need to set up a front company which legitimately deals in cash (e.g., a loan company, pawn shop, etc.) in order to make the criminal activity less conspicuous.

The criminals could move the stolen goods to their private warehouse if they wish.[7] Of course, why maintain such costly and fixed

---

[6] Hoffer (2010, p. 4).

[7] If the conveyance itself is stolen, the criminals would need to move it very quickly since a tracking system could be embedded. Another option is to deliberately set the conveyance along the roadside some distance away from the scene of the crime. By waiting and observing from a distance the criminals can determine whether or not the conveyance is being tracked by the owners and law enforcement.

infrastructure? Why make it easy for law enforcement to stake out their operation? They could use a public warehouse instead to store the goods in the name of the front company.[8] Once they are in storage, the front company could contract with a legitimate for-hire carrier in order to move the goods to where the criminals want them. It is important to note that, apart from the stolen goods, the other transactions appear legitimate: (1) a storage contract between the "owner" of the goods and the public warehouse; and (2) a bill of lading outlining the pick-up, transport, and drop-off of the shipment between the public warehouse and the criminal's preferred destination.

So, the cargo was stolen in transit. What is to become of the original, legitimate shipper whose shipment was stolen and the original, for-hire carrier whose reputation for safety and security is now in question? Well, consider the cargo insurance the for-hire carrier holds. Ideally, then, the insurance company would compensate the shipper once the carrier puts in a claim of cargo theft. However, cargo insurance often does not cover all theft (especially in the cases of unguarded vehicles and unguarded lots).[9] It is up to the carrier to have a more generous policy, or at least be aware of insurance limitations. Worse, what if the driver of the vehicle was in on the theft? Perhaps the driver was bribed to look the other way at a rest stop or refueling point. Since employee infidelity is not covered in cargo insurance, this makes the carrier fully liable on its own to the shipper. Even so, the carrier may have language in the bill of lading that limits liability to some flat amount based on a US statute or based on some lower amount as set by law in a foreign country of origin (in the case of an international shipment).

Consider the free on board (FOB) point which is a term that refers to where ownership passes from seller to buyer. Suppose the FOB point is the seller's loading dock. When a for-hire carrier arrives to transport the shipment from seller to buyer, the sale of the items in shipment

---

[8] Consider the size of the US public warehouse industry. Bond (2015) did a survey of the top 20 public warehouses. These companies control almost 3300 locales across the country comprising around 610 million square feet of storage space. http://www.mmh.com/article/top_20_3pl_and_public_refrigerated_warehouses_2015.

[9] See Seaton (2003, p. 66).

would be consummated. The seller's financial accounts see an increase in revenue on the income statement and the balance sheet will see a fall in inventory balanced against an increase in cash or accounts receivable. The buyer incurs cost, takes on inventory and loses cash or takes on an accounts payable. Well, eventually the buyer may not consider the inventory in transit as true inventory until it is actually received. From a financial accounting perspective we can see why items in transit may truly be in a "black box." Of course, the seller will want the shipment off of his property since he may not wish to have to store and maintain shipments, containers, and conveyances it does not own. The buyer, on the other hand, may not wish the shipment to arrive very promptly due, say, to congestion at his dock or because he is content to maintain an account payable and delay receipt of the shipment. Another possibility is that the buyer wants the item by start of business on Monday while the seller has the item ready to go on Friday afternoon. Unless the origin and destination are more than a 1000 miles apart, the shipment is going to travel on the weekend and will have to be parked in one or more places in order to slow the delivery.[10] Either way, it creates a scenario for the carrier whereby the container or trailer will be parked somewhere for an extended period of time and this increases the risk of theft.

As can be seen FOB_{origin} is an advantageous item for a seller to include in a contract. Naturally, the buyer would prefer $FOB_{destination}$. Of course, we must also consider the carrier. Since a bill of lading is a contract of carriage, like any contract, it requires a meeting of the minds; but it is also up to the shipper (be he the buyer or the seller) to realize that a carrier does not want to expose itself unnecessarily given the limitations imposed by its own cargo insurer. Therefore, it makes sense for shippers and carriers to work in tandem on a cargo security plan to fill in the gaps and loopholes noted above. The question is one of risk mitigation across the supply chain. Which entity is better able to accept the risk?

---

[10] One should not be surprised to learn that weekends have a high concentration of cargo theft. Cole (2016) reports data provided by *CargoNet* which analyzed 3 years of data during the Memorial Day weekend. It was found that 48% of all theft occurred from Friday through Sunday. Focusing solely on electronic shipments, Burges (2013, Fig. 8.5) shows that the increased frequency of theft is also over the weekend.

Maritime Piracy

Piracy is a special form of theft occurring on the high seas or, more technically, international waters. It is estimated to add $15–25 billion per year to global trade costs.[11] A good definition of piracy is provided in Article 101 of the United Nations Convention on the Law of the Sea (UNCLOS, 1982).[12] It includes any of the following acts:

1. any illegal acts of violence or detention, or any act of depredation, committed for private ends by the crew or the passengers of a private ship or a private aircraft, and directed:
   a. on the high seas, against another ship or aircraft, or against persons or property on board such ship or aircraft;
   b. against a ship, aircraft, persons or property in a place outside the jurisdiction of any State;
2. any act of voluntary participation in the operation of a ship or of an aircraft with knowledge of facts making it a pirate ship or aircraft;
3. any act of inciting or of intentionally facilitating an act described in subparagraph (a) or (b).

Article 102 notes that any mutinied crew of a warship, government vessel, or aircraft which engages in the types of piracy noted above will see their vessel/aircraft treated as a private one (meaning it forfeits any immunity conferred by a government on its own infrastructure). In fact, if domestic law allows, the state may declare a pirated vessel/aircraft to be stateless; however, it is typical for the vessel/aircraft to maintain its nationality while under criminal control. Article 100 directs signatories to UNCLOS to cooperate and repress piracy throughout the world. This cooperation is exemplified by Article 105, which deals with seizure of a pirated vessel/aircraft:

> On the high seas, or in any other place outside the jurisdiction of any State, every State may seize a pirate ship or aircraft, or a ship or aircraft taken by piracy and under the control of pirates, and arrest the persons and seize the property on board. The courts of the State which carried out the seizure may decide upon the penalties to be imposed, and may also determine the action to be taken with regard

---

[11] See Jones (2014, p. 158).

[12] It should be noted that the United States is not a signatory to UNCLOS. This was due to objections over the language in Part XI relating to resource and mineral use beyond a country's Exclusive Economic Zone.

to the ships, aircraft or property, subject to the rights of third parties acting in good faith.

United Nations, 1982. United nations convention on the law of the sea. Division for Ocean Affairs and the Law of the Sea. http://www. un.org/depts/los/convention_agreements/convention_overview_ convention.htm.

Thus, while a commercial vessel at sea must fly the flag of a nation state, it enjoys the protection of that state's navy (if it has a viable one).[13] Conversely, a pirated ship may be seized by any navy, military aircraft, or police force (Article 107) in furtherance of justice.

It should be no surprise that a lot of the piracy is concentrated around coastal states which are in turmoil or within some of the marine bottlenecks noted in Chapter 1. The International Maritime Bureau (IMB), which is a division of the International Chamber of Commerce, maintains a database of reported piracy incidents. Its IMB Piracy Reporting Center is an independent nongovernment organization which gathers data from shipmasters around the world. It uses the definition of piracy as set out in UNCLOS. Based on this, Table 3.1 provides a breakdown of incidents.

Given the reported locales of the incidents it is possible to create a map which lays out the danger areas. Fig. 3.1 shows the concentrations of incidents for 2014.

Table 3.1 Maritime Piracy (2014)

Vessel Incidents	Incidents per Region	Violence
183 boarded	5 in South America	266 crews taken hostage
28 attempted attacks	55 in Africa	14 assaults
23 fired upon	34 in the Indian Subcontinent	13 kidnappings
21 hijacked	141 in South East Asia	10 injured
	8 in East Asia	1 killed

Source: IMB Piracy Reporting Center. https://www.icc-ccs.org/piracy-reporting-centre.

---

[13] Most vessel tonnage in the world flies so-called flags of convenience. In this case the commercial vessel wishes to take advantage of the state's easier laws governing labor, wages, taxation, etc. For an overview of flags of convenience see Prokop (2014, pp. 75–79).

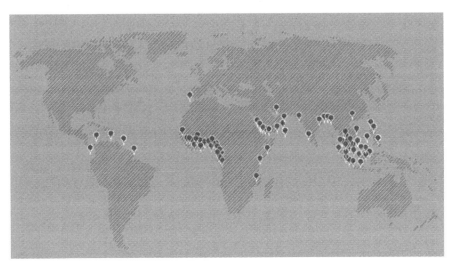

FIGURE 3.1

Maritime piracy concentrations (2014). *Source: Derived from the IMB Piracy Reporting Center.*
https://www.icc-ccs.org/piracy-reporting-centre.

One can see that incidents are concentrated around coastal failed states in sub-Saharan Africa and northern South America, as well as various global choke points. Proceeding from west to east these are: the Strait of Gibraltar, Bab el Mandeb, the Strait of Hormuz, the Strait of Malacca, and several more around the Indonesia and Philippines archipelagos.

Unlike cargo theft, pirates are in it for the ransoms rather than the cargo itself. Of the various types of ocean vessels all are targeted with varying frequencies that not necessarily commensurate with their share of the world fleet. Bulk and container ships have higher piracy shares relative to their fleet sizes. Tanker and chemical ships have piracy shares notably lower than their fleet sizes. General cargo and liquefied gas ships are consistent in their piracy and fleet shares. Table 3.2 gives the incident breakdown by vessel type over the years 1996–2005.

As we can see the pirates do not have clear-cut cargo targets. Like surface cargo theft discussed earlier these ships are targets of opportunity. In 2012, the total ransoms paid amounted to about $90 million which was down from a high of $238 million in 2010.[14]

---

[14] Jones (2014, p. 161) provides a further breakdown of these numbers.

Table 3.2 Piracy and Vessel Type (1996–2005)

Vessel Type	Share in World Fleet	Share in Piracy
Bulk carrier	17.06	28.51
General cargo	30.53	25.16
Container ship	7.16	16.08
Tanker	20.53	15.70
Chemical	21.67	10.58
Liquefied gas	3.05	3.96

Source: Jones, S., 2014. Maritime piracy and the cost of world trade. Competitiveness Review 24 (3), p. 165.

Apart from ransoms, the other costs of piracy include: higher insurance costs (including insurance against wars and kidnapping/ransom), rerouting ships outside of danger zones, traveling at increased speed, and naval protection. All of these costs can be very large. Other social costs include search, capture, and prosecution of the pirates by the countries affected. Estimates of the total costs of all these activities range from $7–12 billion annually.[15]

Terrorism

As opposed to theft, the goal of terrorism is to see trade and transport stopped since it is an important part of a peaceful and functioning society. Terrorists wish to damage the economy of the targeted society and the goods it produces. It is a matter of destroying, not stealing. It is also politically motivated violence and fits somewhere in-between a personal grievance and a state-sanctioned military action. Of course, this raises the question concerning how to deal with terrorists, especially those groups with global reach and revolutionary aspirations.[16] Should terrorism be treated as a law enforcement problem or as a military problem? In other words, are terrorists more criminal or more soldier? This is a policy challenge for a targeted society. Law enforcement would confer rights upon captured terrorists beyond those expected for enemy combatants. On the other hand, if they are enemy combatants

---

[15] Jones (2014, p. 168).

[16] For a complete list and assessment of terrorist groups and the countries where they are located, see Bureau of Counterterrorism (2016).

the country has to recognize that it is at war, and it must follow the international rules of war in dealing with them.

The size of a group large enough to be called terrorist is, of course, quite flexible. What is more definite is that they are held together by some ideology which allows for violence against a targeted society. A good definition of terrorism is found in the United States Code [22 USC. § 2656f(d)(2)]: "[P]remeditated, politically motivated violence perpetrated against noncombatant targets by subnational groups or clandestine agents." Another difference between criminals and terrorists is that the latter usually claim to be committing their acts in the name of freedom. In other words, they feel they are being repressed in some way by the government and society they target. However, they usually attack civilians to get the attention of that government.

Why would a terrorist target a supply chain and/or the logistical flows that sustain it? Basically, such a target in our interconnected world comports with many of their goals. Consider why the terrorist threat seems to be inordinately focused on the aviation industry.

- Commercial aviation is a symbol of modern mobility and leisure
- Multiple countries form the interconnected network of global commercial flights
- Passenger flights and courier air cargo are a fundamental feature of modern economic activity
- Disasters in the air or at airports are given global media coverage

This means that aviation is a tempting target because the consequences can be enormous for the targeted country and for global trade. If the mission to take down even a single aircraft with just a few innocent civilians on board is successful, the whole world could be watching the aftermath in real-time both on television and the Internet.

Terrorists and common thieves who target supply chains share a trait of conservatism. Each has limited resources combined with the need to be successful. But this is also combined with a trait of opportunism meaning each is willing to risk failure and the loss of their resources if the chance for success is high enough to them. This means they have to plan carefully, have a lot of patience, and test for weak points along the supply chain. History has not been on the terrorist's side to this point since no supply chain has been disrupted for more than a few days (with 9/11 being a prominent example of restricting public air space and cutting land border crossings into the United States to a trickle). However, there is no doubt that terrorist groups are seeking a high-profile, symbolic target which if hit would cripple the economy.

The terrorist's arsenal has grown sharply in the twentieth century as bomb-making became more sophisticated and access to automatic and semiautomatic weaponry has increased. All of this has combined with the older, and less prevalent, practice of suicide missions to create a lethal threat to targeted areas. The drama of 9/11, with suicide bombers essentially using airplanes as guided missiles, seared this intent and extent of terrorists to plan and carry out their missions in the minds of all who watched it on television. The Internet has allowed information and propaganda to spread at rapid speed and in the quasi-personal way that all of us feel connected to the cyber-world.

The challenge law enforcement faces in dealing with the modern terrorist threat involves the linking of three trends since 9/11:

- The groups are more decentralized and made up of lone wolves
- The Internet is being used for spreading propaganda and for financing activities
- Terrorists are organized into international crime groups

Military and police operations can defeat terrorists groups in battle. However, this victory can also simply splinter and scatter the combatants. The war continues because the terrorists' ideology is not yet defeated. This means that cyberwar over the Internet is necessary to counter the propaganda and disrupt the movement of information and funds. This, of course, requires a multinational effort. Apart from terror, many groups resort to the standard activities of organized crime to finance their larger goals. These activities include theft, ransom, drug trafficking, and money laundering.

The threat can seem existential when considering the range of weapons seemingly at the terrorist's disposal. Chemical, biological, radiological, nuclear, and explosive (CBRNE) describes the set and the popular term is "weapons of mass destruction." Of course, this is more aspirational on the part of terrorists than actual. Nevertheless, the public has deep concern over the possibilities of CBRNE reaching the hands of a terrorist willing to die in order to accomplish his mission. One could also include cyber-terrorism in the form of hacking into infrastructure controls and creating havoc.

Counterfeiting

Counterfeiting legitimate items is a form of intellectual theft. To the extent that counterfeit items mislead the buyer, they infiltrate legitimate supply chains and dilute any intellectual property rights (IPRs) held by any and all producers and vendors along this supply chain. Even if the buyer is aware that the item is fake (e.g., a street vendor selling $50 "Rolex" wristwatches in Hong Kong), he is serving to keep an illegitimate supply chain in business which will still dilute IPRs if other buyers are misled. Counterfeit items include apparel, electronics, medical devices, pharmaceuticals, and automobile parts. With the increase in world trade and global supply chains to facilitate it, there is a temptation for counterfeiters in poor or emerging markets to leverage the demand for, and reputation of, high value-added products. The overall results are:

- Higher prices for legitimate products since the producers must spend time and money on improved packaging, transportation, and litigation.
- Government earns less tax revenue. Also, illegitimate supply chains, as a means for organized crime and terrorism to take root, pose a national security risk.
- Strained trade relations with countries which cannot or will not recognize IPRs.

The operational and financial scope of counterfeiting is very wide. Since the counterfeits are lower quality, they can impact the health and safety of the user. Over the years 2004–2009, US Customs and Border Protection seized $1.1 billion worth of counterfeit goods. Of these the largest shares were: footwear (32%), wearing apparel (15%), handbags/wallets (11%), consumer electronics (9%), computers (5%), and pharmaceuticals (5%). One country accounted for the majority of these imports: China's share was 77%.[17] Regarding pharmaceuticals, "The World Health Organization (WHO) estimates that as much as 10 percent of the global pharmaceutical market—a half-trillion-dollar marketplace—is counterfeit. In some countries, the WHO estimates that 25 percent or more of the entire drug supply is counterfeit … The Federal Bureau of Investigation estimates that the financial impact of counterfeit drugs on US companies is $30 billion a year."[18]

---

[17] U.S. Government Accountability Office (2010, pp. 7–8).
[18] Wyld (2008, p. 206).

In order to mislead the buyer the counterfeit item has to enter either the legitimate supply chain or an illegitimate one. The counterfeiter can do this by:

- Mixing counterfeit and legitimate versions of the product in the same shipment to be sent to the customer along an authorized distribution channel.
- Shipping a dedicated shipment of counterfeit items unassembled in order to make it harder for the customer to recognize them. This distribution channel is still authorized but has been infiltrated.
- Creating a parallel (though illegitimate) supply chain by stealing the identity of the legitimate producer in order to sell the counterfeit items directly to wholesalers or retailers. For imported items, the counterfeiter could steal the identity of the importer in order to get the items cleared through customs, and then sell the items along unauthorized distribution channels.

Inadvertently buying a fake Gucci purse, for example, will inconvenience a consumer. However, a fake medical device or automobile part can have more dire consequences. These items may fail prematurely and may not be subject to the legitimate supplier's warranty protection. Therefore, it is necessary to devote some time and money to deterring and detecting counterfeits. One way to deter is to buy only from vendors which have the necessary trademarks on the finished goods or spare parts. Failing the ability to buy from original equipment manufacturers (OEMs), one could buy from OEM-authorized distributors. Vetting can take place through trade associations or certification bodies. Detection of counterfeits can be handled through statistical quality control; that is, pulling parts at random and authenticating them through serial numbers or special imprints and holograms. In all cases, when the buyer and vendor complete the transaction and a transportation carrier is now in the picture all three entities should establish a chain of custody for the parts or goods to be moved. The motto should be: test it if you cannot trace it. Finally, firms and their vendor partners should consider joining the Government-Industry Data Exchange Program, which is a cooperative between industry and government in order to reduce resource costs by sharing technical information.

Corruption and Maverick Buying

Corruption can impact supply chain security in that it makes input purchasing and cargo transport less predictable in terms of both time and cost. More time on the move means more opportunities for things to go wrong; and more cost means there are less funds to devote to running the business. Paying bribes is the hallmark of corruption. Corruption involves an abuse of authority for the personal gain of the person(s) trusted with that authority. The payoff may come in the form of a bribe (which is somewhat voluntary) or it may be extorted (which is involuntary). When it comes to supply chain management (and indeed all types of management) ideally we would like the process to be as transparent as possible; but corruption introduces an element which comes at the expense of complying with rules and regulations.

Corruption can take place in the private sector and the public sector. Carriers could abuse their power to the detriment of a shipper, and government officials (e.g., customs officers) can do likewise to shippers and carriers. Such activity is more prevalent in less developed countries where the rule of law may be less widespread. Furthermore, it is often easier for such entities to take advantage of foreigners who may not be able to, or even wish to, try to fight off demands for bribes by seeking recourse in unfamiliar court systems.

Of course, it must be admitted that sometimes people are happy to pay bribes. Being realists, they see these payments as "grease money" used to simply insure that people do their job and move things along. Think about the congested ports in the emerging economies in the Far East and how a US shipper might be happy to pay "extra" to insure his items leave in a timely manner. Legally, this is a tricky area because one must distinguish between payments for services that were to be performed anyway versus payment to induce a service.[19]

Whether intentional or not maverick buying involves the breaking of established purchasing contracts for personal reasons. It is an internal

---

[19] The U.S. Foreign Corrupt Practices Act (FCPA) of 1977 prohibits bribery but appears to be more accommodating of grease money. The FCPA was amended through the passage of the International Anti-Bribery and Fair Competition Act of 1998, which is meant to comport with international law; specifically, that of the Organization for Economic Cooperation and Development (OECD). In 1989, the OECD passed the Convention on Combating Bribery of Foreign Public Officials in International Business Transactions.

gray market—a dark supply chain alongside the legitimate one. To the extent that these fall outside of supply security plans it can open the firm to vulnerabilities, make items harder to track, and harder to measure success. The market is gray as opposed to black because the act is not criminal, but it does impose costs on the supply chain. Improved security could mitigate the problem through vendor-specific purchasing cards, tighter electronic purchasing controls, etc. Discouraging maverick buying is also a matter of organizational change and insuring buy-in down the hierarchy of employees when upper management is considering switching from vendor A to vendor B.[20]

NATURAL THREATS

Weather Vulnerabilities

Supply chains are spread out over land, water, and air. Logistics must traverse topographical spaces. Therefore, these activities are vulnerable to the forces of nature and unpredictable acts of God.[21] In other words, one cannot forget that all business activities are dependent on the natural environment in which they take place.

Floods can happen in all 50 states of the United States and it does not take much water on the ground before damage becomes significant. Just a few inches can cause property damage while just a couple of feet can wash away automobiles. As an example of a tracking map, Fig. 3.2 shows the extent of the flood risk in the United States for 2010. The map reflected an El Nino effect in the south and East Coast, as well as an above average snowpack in the upper Midwest.

The NOAA website provides an up-to-date detailed map on changing flood risks, as well as other natural hazards which are discussed in this section.[22]

Floods can be by-products of hurricanes which is a seasonal phenomenon (around June through November in the United States) and can hit coastal states from the Gulf of Mexico northward to

---

[20] For a complete review, see Karjalainen et al. (2009).
[21] Bullock et al. (2013). Chapter 3 provides a more detailed overview of a variety of hazards which can negatively affect the supply chain.
[22] See NOAA's National Weather Service map at: http://www.nws.noaa.gov/view/largemap.php.

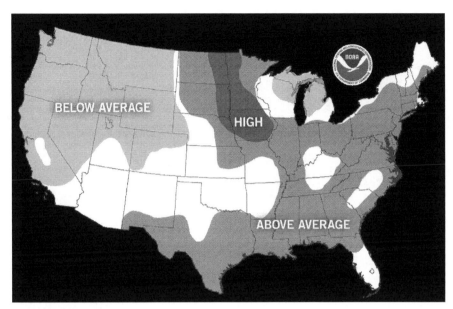

FIGURE 3.2

Flood vulnerability (2010). *Source: National Oceanic and Atmospheric Administration (NOAA). U.S. Department of Commerce.*

New England. Not only is most of the country's populace concentrated along the vulnerable East Coast but some of the most complex and expensive business activities are located there (for example, the dozens of crude oil platforms along the Gulf of Mexico coast and miles offshore). By contrast, the entire West Coast from the Aleutian Islands in Alaska southward to Mexico are vulnerable to earthquakes. Unlike hurricanes, earthquakes arrive with little warning, can damage any anchored structure, and can cause tsunami waves when the epicentres are situated near shore or offshore. Fig. 3.3 shows where the risks are highest.

The Midwest from Texas northward to South Dakota are vulnerable to tornadoes. While not as sudden as earthquakes, and yet not as predictable and slow moving as hurricanes, tornadoes are a function of storm systems caused by the mixing of cold and warm air masses along the Prairies. The very strong winds of a tornado can create damage along the same lines as hurricanes except that there is little or no strong rain accompanying. Fig. 3.4 shows where tornadoes are situated.

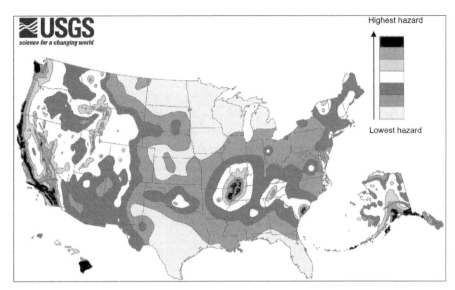

FIGURE 3.3

Earthquake vulnerability (2014). *Source: U.S. Geological Survey. Earthquake Hazards Program.*

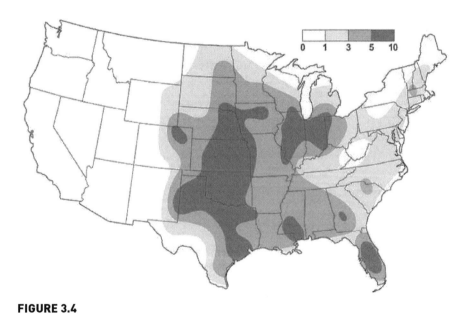

FIGURE 3.4

Annual Tornado Reports per 10,000 square miles (1950–1995). *Source: National Oceanic and Atmospheric Administration (NOAA). U.S. Department of Commerce.*

Seasonal extremes can lead to a destructive chain of events as well. Hot, dry summers can trigger wildfires in forested or even shrubby areas. These summers can also cause droughts. On the other hand, snowy winters can lead to flooding in the spring. Thunderstorms can bring about lightning strikes which can cause fires on the ground or aircraft failures in the sky. They can also cause hail damage to fixed structures and vehicles. Finally, leaving aside whether or not humans are a major cause of climate change, the apparent long–term cycle of climate change represents another challenge. Coastal erosion, disappearing wetlands, Arctic ice melt, and increased acidity in ocean waters will have effects on fish, shell fish, and plants. Warmer temperatures will affect crop cycles and alter the distribution of arable land across the world.

All of these vulnerabilities make it imperative for supply chain partners to maintain emergency management plans on top of the services federal, state, and local governments may provide in times of emergency.

Hazardous Material (Hazmat)

Hazardous materials (hazmat; also referred to as dangerous goods) would not exist unless there was a demand for them (e.g., gasoline, radioactive isotopes, etc.). They would not exist unless they were a necessary by-product of a production process (e.g., toxic waste). As such, these items need to be transported from one point to another as a matter of satisfying a market (in the case of gasoline) or being environmentally responsible (in the case of disposing of toxic waste). Transacting to transport these items takes place in much the same way as other cargo except that the shipper and the carrier face a higher level of care and responsibility.

In 2007, the US transportation system carried 2.2 billion tons of hazmat generating 324 billion ton-miles of activity.[23] This material was valued at $1.4 trillion. More than half of this tonnage (about 58%) was transported on the highways (meaning by motor carrier). Of the 2.2 billion tons of hazmat, the most prevalent commodity (1.8 billion tons worth) was flammable liquids, mostly refined petroleum products (moved by all transport modes except air). Pipeline had a 27% tonnage share followed by water and rail (both tied at 5%).

---

[23] National Transportation Statistics Annual Report (2016). Table 1-62.

The U.S. Department of Transportation (USDOT) defines nine classes of hazmat. These are:

- Class 1 (Explosives)
- Class 2 (Gases)
- Class 3 (Flammable liquids)
- Class 4 (Flammable solids)
- Class 5 (Oxidizers and organic peroxides)
- Class 6 (Toxic materials and infectious substances)
- Class 7 (Radioactive materials)
- Class 8 (Corrosive materials)
- Class 9 (Miscellaneous dangerous goods)

In terms of transport mode each has its own standards and practices. Different states may have individual requirements on top of federal ones. At the international level the most prevalent modes are air and ocean vessel, so it is not surprising that each has its own published standards. The UN's International Maritime Dangerous Goods Code was adopted in 1965 and contains the same hazmat classes which USDOT uses. The International Civil Aviation Organization (ICAO) publishes under Annex 18, Technical Instructions for the Safe Transport of Dangerous Goods by Air (simply known as the "Technical Instructions"). Naturally, these rules differentiate between commercial passenger and commercial cargo aircraft. "There is generally no restriction on the number of packages per aircraft. The Technical Instructions also give the packing methods to be used and the packagings [*sic*] permitted, together with the specifications for those packagings [*sic*] and the stringent testing regime that must be followed. There are also requirements for the markings and labels for packages and the documentation for consignments."[24]

In the simplest sense dealing with hazmat involves following the rules and regulations as set down by the jurisdiction(s) to be traveled. Apart from training and certifying operators, hazmat transport requires careful attention to determining classification, packaging, marking/labeling, and documenting activity. Finally, the carrier should make sure the authorities are aware of the origin/destination/route/etc. of the transport so as to facilitate emergency response should it be necessary.

---

[24] International Civil Aviation Organization (ICAO) (2012).

Infrastructure Vulnerabilities

Man-made structures are vulnerable to the weather conditions noted above. They are also vulnerable to simple wear and tear and to inferior design. Wiring problems can lead to fires; excess water can cause dams to fail causing flooding; improper maintenance can cause nuclear power plants to meltdown; careless transportation of hazardous material can lead to environmental spillage. An overloaded power grid can cause blackouts over wide ranging areas.

While government regulation and insurance can mitigate some of the infrastructure risk and risk-taking that private sector organizations might tolerate, the status of the public infrastructure is a different matter. In this case, risk-tolerance and conditions are matters of politics; that is, the amount of taxes the public will tolerate and how much of it they wish to see devoted to maintenance. The public sector saw massive growth in roads, bridges, ports, and airports in the decades after World War II. Once built it is a matter of maintaining them. However, fiscal and political realities have made it progressively harder to generously fund improvement projects.

The Federal Highway Administration collects data on road conditions. For example, in 2013, of the approximately 726,000 miles of road, 74,000 miles were classified as mediocre and 66,000 were classified as poor. Combined, this implies that 19% of the roads are in less than adequate condition.[25] Airport runways are much better with only 17% characterized as fair and 2% as poor.[26] In terms of bridges, about 10% are structurally deficient and 14% are functionally obsolete.[27] These numbers are expected to rise as years pass without adequate infrastructure funding.

ISSUES AND PROBLEMS GOING FORWARD

Shippers and carriers, including their intermediaries, need to work together in order to mitigate the risk of theft. All of them can be a target. The role of government becomes crucial because it sets up the laws and regulations the shippers, carriers, and intermediaries must comply

---

[25] The measure is called the International Roughness Index and a complete state by state breakdown may be found in State Transportation Statistics (2015). Table 1-4.
[26] National Transportation Statistics Annual Report (2015). Table 1-25.
[27] State Transportation Statistics (2015). Table 1-7.

with when attempting to deter theft. Dealing with natural threats is also a government responsibility. The role of government is the focus of Chapter 4. When considering the motivations of thieves and terrorists it is worthwhile to consider whether the government program to deal with them is proactive or reactive. Each has its attributes. Of course, cyberspace is also a front for crime and terrorism. Chapter 5 examines the threats and security measures involved there. How the private sector works with the government is the focus of Chapter 6. The interactions of all of these players is a deciding factor on the success or failure of security programs, laws, and regulations.

Bibliography

Bond, J., December 1, 2015. Top 20 3PL and Public Refrigerated Warehouses, 2015. Modern Materials Handling. http://www.mmh.com/article/top_20_3pl_and_public_refrigerated_warehouses_2015.

Bullock, J.A., Haddow, G.D., Coppola, D.P., 2013. Introduction to Homeland Security, fourth ed. Butterworth-Heinemann, Waltham, MA.

Bureau of Counterterrorism, 2016. Country Reports of Terrorism 2015. United States Department of State, Washington, DC.

Burges, D., 2013. Cargo Theft, Loss Prevention, and Supply Chain Security. Butterworth-Heinemann, Waltham, MA.

Cole, M., May 24, 2016. Cargo Theft Firms Warn Truckers of Increased Theft Activity During Memorial Day Weekend. Overdrive. http://www.overdriveonline.com/cargo-theft-firms-warn-truckers-of-increased-theft-activity-during-memorial-day-weekend/.

Federal Bureau of Investigation, 2010. Inside Cargo Theft: A Growing, Multi-Billion Dollar Problem. Washington, DC https://www.fbi.gov/news/stories/2010/november/cargo_111210/cargo_111210.

Federal Bureau of Investigation, 2015. Uniform Crime Reporting Program: Cargo Theft Update. Washington, DC https://www.fbi.gov/about-us/cjis/ucr/cargo-theft-update.

Fischer, R.J., Halibozek, E.P., Walters, D.C., 2013. Introduction to Security, ninth ed. Butterworth-Heinemann, Waltham, MA.

Hoffer, E., 2010. The mechanics of supply chain theft. In: Thomas, A.R. (Ed.), Supply Chain Security: International Practices and Innovations in Moving Goods Safely and Efficiently, vol. 1. Praeger, Santa Barbara, CA, pp. 1–51.

IMB Piracy Reporting Center. https://www.icc-ccs.org/piracy-reporting-centre.

International Civil Aviation Organization (ICAO), 2012. The Transport of Dangerous Goods by Air. http://www.icao.int/safety/DangerousGoods/Pages/background.aspx.

Jones, S., 2014. Maritime piracy and the cost of world trade. Competitiveness Review 24 (3), 158–170.

Karjalainen, K., Kempainen, K., van Raaij, E.M., 2009. Non-compliant work behaviour [sic] in purchasing: an exploration of reasons behind maverick buying. Journal of Business Ethics 85 (2), 245–261.

LoJack, S.C.I., 2015. 2015 1st Quarter Supply Chain ISAC Report of Cargo Theft Activity. SC-integrity, Inc., Richardson, TX. http://www.lojacksci.com/?wpfb_dl=33.

McNicholas, M., 2008. Maritime Security: An Introduction. Butterworth-Heinemann, Waltham, MA.

National Oceanic and Atmospheric Administration (NOAA). U.S. Department of Commerce.

National Transportation Statistics Annual Report, 2016. Bureau of Transportation Statistics, Washington, DC.

National Transportation Statistics Annual Report, 2015. Bureau of Transportation Statistics, Washington, DC.

Prokop, D., 2014. International transportation management. In: Prokop, D. (Ed.), The Business of Transportation. Applications, vol. 2. Praeger, Santa Barbara, CA, pp. 70–89.

Seaton, H.E., 2003. Protecting Motor Carrier Interests in Contracts, second ed. VA. Seaton & Husk, L.P., Vienna.

State Transportation Statistics, 2015. Bureau of Transportation Statistics, Washington, DC.

United Nations, 1982. United nations convention on the law of the sea. Division for Ocean Affairs and the Law of the Sea. http://www.un.org/depts/los/convention_agreements/convention_overview_convention.htm.

U.S. Geological Survey. Earthquake Hazards Program. http://earthquake.usgs.gov/hazards/products/conterminous/.

U.S. Government Accountability Office, 2010. Intellectual Property: Observations on Efforts to Quantify the Economic Effects of Counterfeit and Pirated Goods. U.S. Government Accountability Office, Washington, DC.

World Bank, 2014. World Data. http://data.worldbank.org/region/WLD.

Wyld, D.C., 2008. Genuine Medicine? Why safeguarding the pharmaceutical supply chain from counterfeit drugs with RFID is vital for protecting public health and the health of the pharmaceutical industry. Competitiveness Review 18 (3), 206–216.

CHAPTER 4

The Role of Government

CONTENTS

SUPPLY CHAIN SECURITY PROGRAMS

The Basics

Whether it is a private business wishing to enhance security in order to protect its reputation, profits, etc. or it is a government wishing to regulate the security programs of a private business, it is fairly straightforward to outline what a basic security plan should look like.

65

Global Supply Chain Security and Management. http://dx.doi.org/10.1016/B978-0-12-800748-8.00004-2

- Define the activity: the item, the storage, the transport, the names, and locales of the parties involved. Ideally, all parties along the supply chain should have a hand in building the plan in order to secure buy-in.
- Define the threats: mitigation and reaction to a security breach is easier when one knows what to expect.
- All those handling the items in storage or in transport should be properly licensed and have proper levels of training and experience. Proper identification must always be carried and numerously checked.
- Access to items in storage or in transport should be limited only to what is necessary to perform the job.
- Incoming cargo must be screened for security before arrival using a protocol. There should also be a protocol for the possibility of physical inspection of any suspicious cargo before departure or, at least, upon arrival.
- Outgoing cargo contents should be validated for accuracy of the transaction (i.e., price, quantity, quality, and documents).
- Locks, seals, and some form of tracking program should be in place while the items are in transport. Items in storage need to be within a secured facility.
- Data related to the items above should be gathered.
- Performance should be measured against best practices.
- Risk should be mitigated by performing regular audits, practice drills, and having random inspections.
- Develop relationships with local, state, and federal law enforcement.

It is important to note that the points above represent just the basics and are not meant to imply that a cookie-cutter approach can be taken when it comes to supply chain security planning. Customization is necessary. Complexities in these plans occur when we differentiate among shipment types, modes of transport, vendors, and ports of entry. Another source of complexity occurs because different arms of the government may require different sets of regulations to comply with. As well, the amount of data gathered could be enormous depending upon the volume of activity. Making sense of all the data, let alone making decisions based on it, can be challenging. But what is necessary is to have a basic plan that requires training, documentation, and tracking. If a breach in security does occur, the plan has to have response mechanisms in place and protocols for post-breach investigations. Finally, the plan must be reevaluated to see what can be learned from the breach.

THE DOMESTIC CONTEXT

An area of concern for the federal government is commercial air cargo transported along with passengers. In this case, the Transportation Security Administration (TSA), a part of the Department of Homeland Security (DHS), is concerned with domestic cargo when it travels with passengers. As will be discussed below, it is Customs and Border Protection (CBP) which oversees inbound international air cargo.

A known shipper is one who is known to the carrier (and, through mandated reporting, known to the federal government) because of the previous business they had. For a first-time shipper to become "known" a carrier has to physically visit the premises so that the carrier could legitimately vouch for the shipper. Items to look for are a chain of custody of inputs into, and output from, the shipper's premises. There should also be no evidence of smuggling activity (with contraband mixed in with legitimate items). Notice how this known shipper classification sets up a two-tiered class of shippers meaning that one group should be given more scrutiny by authorities than another.

In air cargo, the TSA maintains a database of known shippers. The intent is to make sure that cargo carried on passenger airplanes has been screened since passengers are themselves screened. Any package weighing more than one pound must be from a known shipper in order to be carried along with passengers. All heavier cargo from unknown shippers would have to be shipped by air cargo freighters. In recent years the TSA has been criticized for not providing clear guidance for carriers when visiting their shipper customers.[1] There are currently about 1.5 million shippers listed in the known shipper database.

THE INTERNATIONAL CONTEXT

What happens when an international border crosses one or more points along the supply chain? Quite a lot. CBP enters the picture

---

[1] Edmonson (2009, p. 15).

as the protector of the ports of entry, the prime regulator of international trade, and the collector of tariffs and duties. CBP is concerned with international cargo security; that is, inbound cargo from over 100 trade partner countries arriving by all modes of transport. Any imports arriving at a port of entry (i.e., border crossing, airport, or ocean vessel port) are under the custody of CBP unless and until they are released and allowed to become part of the US economy. An importer may deal directly with CBP or may use the services of a licensed customs broker to act as an intermediary.

Automated Commercial Environment

The Automated Commercial Environment (ACE) has taken almost 20 years of planning and coordination on the part of CBP. The target date for full implementation is 2017. ACE uses information technology to allow required documents to be uploaded and payments to be made online and, thereby, speed up customs clearance for imports. This paperless system will be facilitated by a web-based system known as the International Trade Data System. Support documents may be sent in pdf or jpeg formats; in other words, image files as opposed to web-friendly e-forms. ACE will also be used in the exporting process. For imports the legacy system it is replacing is known as the Automated Commercial System. For exports the legacy system is known as the Automated Export System.[2] Customs brokers will also transition to ACE and leave behind their special portal to CBP known as the Automated Broker Interface. Thus, ACE is to provide a "single window" portal for all parties to communicate with CBP.

ACE will also be the means through which importers and exporters will interface with other partner government agencies (PGAs) outside of CBP which regulate or are dependent on trade. The single window will allow trade documents to be shared among the PGAs.[3] If ACE is rolled out

---

[2] One reason the U.S. government regulates exports is for national security reasons and for discouraging trade with nations under trade sanctions. Another reason may be that export licenses are required for the item in question.

[3] As of September 2016 the roll-out of ACE has been three years overdue. ACE has also been about $1 billion over budget. See Hutchins (2016a, p. 72). The PGA which was part of the initial roll-out is the Animal and Plant Health Inspection Service in order to assist compliance with the Lacey Act which prohibits trafficking in illegal wildlife. According to CBP about 60% of cargo imports are being uploaded by shippers of brokers to ACE. With 40% more to comply and dozens more PGAs to make their requirements available for upload and review on ACE, it is still too soon to tell if ACE will be successful in the years to come.

properly this should speed the trade compliance process since all PGAs will see the uploaded information at the same time. Shipment delays are often the result of holds placed on them by agencies other than CBP while the former awaits information from the latter. ACE should mitigate this since it is supposed to function much like enterprise resource planning (ERP) systems used in the private sector. To date there are 48 PGAs (including CBP) as shown in Table 4.1. Currently, these agencies collect over 200 different forms from importers and exporters. ACE is intended to reduce the number to one electronic form with support documents.

ACE will be used for ocean, truck, rail, and air transport modes. The connecting of all these interfaces makes this a very challenging application of electronic data interchange (EDI) technology. Items to be filed by importers and exporters include:

- Importer Security Filing (ISF) data; also known as "10 + 2" (handled by the importer and the ocean vessel carrier). The ISF only applies in the ocean vessel mode. The importer in this case could be: the cargo owner, the cargo purchaser, the consignee, or an intermediary such as a customs broker. The first 10 items are to be provided by the importer and last two by the ocean vessel carrier at least 24 h prior to lading at the foreign port (commonly known as the "24 h rule"):[4]
  - Manufacturer/supplier name and address
  - Seller name and address
  - Buyer name and address
  - Ship-to party's name and address (i.e., the party who will take immediate physical receipt of the cargo when released by CBP)

---

[4] There are different ISF requirements for cargo simply transiting through the United States on the way to a foreign port [i.e., for freight remaining on board (FROB)]. The carrier will be considered the importer of record in this case by CBP and will need to provide five elements to the ISF any time prior to lading at the foreign port of origin. These are: (1) name and address of the booking party (i.e., the entity requesting space on the vessel); (2) cargo's final port of destination; (3) final place of delivery; (4) name and address of ship-to party; and (5) commodity harmonized tariff schedule number written to at least six digits of detail. If the transiting cargo is to be stored and transferred to another vessel in the United States, the ISF will have to be provided 24 h prior to foreign port departure. In either case, the onus for the carrier to maintain strong ties with his shipper customer is increased. Another exception is for breakbulk ocean vessel cargo. In this case the importer and carrier must file the ISF 24 h prior to US arrival instead of foreign port departure.

Table 4.1 Partner Government Agencies in the Automated
Commercial Environment

Department of Agriculture	Agricultural Marketing Service
	Animal and Plant Health Inspection Service
	Food Safety and Inspection Service
	Foreign Agricultural Service
	Grain Inspection, Packers & Stockyards Administration
Department of Commerce	Bureau of Industry and Security
	Census Bureau
	Enforcement and Compliance
	Foreign Trade Zones Board
	National Marine Fisheries Service
	Office of Textiles and Apparel
Department of Defense	Army Corps of Engineers
	Defense Contracts Management Agency
Department of Energy	Energy Information Administration
	Office of Fossil Energy
	Office of General Counsel
Department of Health and Human Services	Centers for Disease Control and Prevention
	Food and Drug Administration
Department of Homeland Security	Coast Guard
	Customs and Border Protection
	Transportation Security Administration
Department of the Interior	Fish and Wildlife Service
Department of Justice	Bureau of Alcohol, Tobacco, Firearms and Explosives
	Drug Enforcement Administration
Department of Labor	Bureau of Labor Statistics
Department of Transportation	Bureau of Transportation Statistics
	Federal Aviation Administration
	Federal Highway Administration
	Federal Motor Carrier Safety Administration
	Maritime Administration
	National Highway Traffic Safety Administration
	Pipeline Hazardous Materials Safety Administration
Department of the Treasury	Alcohol and Tobacco Tax and Trade Bureau
	Financial Crimes Enforcement Network
	Internal Revenue Service
	Office of Foreign Assets Control

Table 4.1 Partner Government Agencies in the Automated Commercial Environment *Continued*

Department of State	Bureau of Administration, Office of Logistics Management
	Bureau of Ocean and International Scientific Affairs
	Directorate of Defense Trade Controls
	Office of Foreign Missions
Independent Government Agencies	Consumer Product Safety Commission
	Environmental Protection Agency
	Federal Communications Commission
	Federal Maritime Commission
	International Trade Commission
	Nuclear Regulatory Commission
	Office of the United States Trade Representative
	U.S. Agency for International Development

Source: U.S. Customs and Border Protection, 2016. ACEopedia, Washington, DC. Publication Number: 0293–1015, p. 5.

- Container stuffing location
- Consolidator/stuffer name and address
- Importer of record number or foreign trade zone applicant identification number [e.g., social security number or employer identification number (EIN)]
- Consignee number(s) [e.g., social security number or employer identification number (EIN)]
- Country of origin of the cargo (i.e., where it was last manufactured, assembled, or grown)
- Commodity's harmonized tariff schedule number to at least six digits[5]

[5] The Harmonized Tariff Schedule of the United States (HTSUS), which determines how to classify imports and assess tariffs/duties, is published by the U.S. International Trade Commission (ITC). The ITC uses international standards set by the World Customs Organization (WCO). The first two digits in a classification refer to a commodity chapter while further digits refer to headings and subheadings. For example, consider an item classified as 6401.92.30. HTSUS Chapter 64 refers to footwear; heading 01 refers to waterproof; subheading 92 refers to footwear covering the ankle but not the knee; and subheading 30 refers to ski boots. All WCO countries require at least six digit reporting though HTSUS can provide detail up to 10 digits.

- Vessel stow plan (i.e., how the various containers are stacked in the vessel)
- Container status messages
- Import Cargo Manifest (handled by the carrier). This will be discussed in greater detail below
- Export Commodity Data (handled by the exporter)
- Any Required Support Documents (handled by importer, exporter, carrier, third-party logistics provider (3PL), or customs broker as appropriate)

Once ACE is running without too many glitches, other government agencies involved in regulating imports [e.g., the Food and Drug Administration and the Environmental Protection Agency (EPA)] will be able to tap into ACE as well.[6] Apart from the time to achieve the smooth operation of the electronic data interchange system, other import operations may take time to integrate. For example, in 2016 CBP raised the duty-free *de minimis* shipment level from $200 to $800. This means that more items can be placed in a single package with no tariffs or duties assessed. This is certainly helpful to small shippers or those with many, but low value, shipments such as e-commerce retailers. Such retailers may wish to take advantage of the savings from the new duty-free level to handle imports on their own (i.e., use a private customs broker and any commercial airline instead of air cargo integrators, especially if their all-inclusive brokerage services are too expensive at the moment). The problem with ACE occurs with the import cargo manifest which is the responsibility of the carrier and summarizes all the cargo on board the conveyance. Since there is no need to file a customs entry for *de minimis* items (and, therefore, no accompanying documents from a customs broker) CBP would have an onerous task of expediting the clearing of such cargo if the conveyance is carrying hundreds of shipments. CBP would have to try to sort out *de minimis* shipments from the others given that all they have is an aggregated import cargo manifest which does not indicate individual values. CBP would have to examine the manifest for the individual shipments' bills of lading reference numbers to determine the individual cargo values. Thus,

---

[6] According to CBP there are 47 other agencies involved in regulating US imports and exports. About 200 forms can be required to comply with all of their requirements for importing and exporting. For a listing of the agencies, see U.S. Customs and Border Protection (2016).

expeditious treatment is not possible unless the broker has the fore-sight to ask the carrier to attach a document to his manifest when the latter uploads it in ACE.

If this "single window" of trade compliance is made to work, it will greatly assist the efforts of importers, carriers, 3PLs, and customs brokers. Indeed, given that the manifests will be uploaded in ACE prior to the cargo's arrival at a port of entry, it might speed the trade process while allowing CBP to enhance its risk management and target illicit shipments. Access to data by all parties is expected to be in near real-time. Cargo entry and cargo release data is to be updated in the ACE portal on a daily basis. Technology may mitigate the trade-off between free flowing trade and supply chain security. If the glitches can be worked out, then the single window will have CBP acting as an intermediary to handle the documents all the other agencies wish to access. Also, the promised status messages will allow importers and exporters to know when cargo is in transit or at a port of entry.

Customs Trade Partnership Against Terrorism

Customs Trade Partnership Against Terrorism (C-TPAT) began in November 2001. Though voluntary it now is the world's largest supply chain security program. It is also a public-private partnership (PPP). CBP's partners (accounting for over 50% of US imports by value) include:

- US importers/exporters
- US/Canada/Mexico motor carriers
- US/Canada/Mexico rail carriers
- Ocean vessel carriers
- US customs brokers
- US ocean port authority/terminal operators
- US freight consolidators/intermediaries
- US/Canada/Mexico manufacturers

Managed by CBP, the partnership is intended to secure cargo beginning at its source. Today more than 11,000 companies have been vetted by CBP and are members of the program. Importers/exporters and carriers make up about 40% each with the other 20% being brokers, intermediaries, port operators, and manufacturers. The incentive to join is the promise of expedited treatment at a US port of entry. During the first few days after

9/11 the US border was shut down to the majority of air, rail, and motor carrier commercial activity. When the border reopened, security measures were stepped-up. However, C-TPAT was intended to be an incentive for shippers (more specifically, US importers and Canadian/Mexican exporters[7]) to deploy industry's best practices. Domestic or foreign (specifically, Canadian and Mexican) carriers, freight forwarders, and brokers could also join in order to deploy their own industry best practices. This changed CBP's role from just policing to include partnering.

CBP views best practices as: "innovative security measures that exceed the C-TPAT minimum security criteria and industry standards ... include high-level managerial support, employ a system of checks and balances, and have written and verifiable policies and procedures."[8] CBP has identified eight areas where some private sector organizations have met with CBP's approval. These are set out in Table 4.2.

Members of C-TPAT sign a Memorandum of Understanding noting that CBP has the right to vet their supply chain security measures and, if necessary, expel them from the program. CBP uses a validation process to evaluate the member's "supply chain security profile." Items which members must include in their profiles relate to securing: inbound and outbound shipments; buildings and surrounding infrastructure (which is certainly daunting for railways); personnel; and conveyances. Finally, members also need to devote time and money to set up training and awareness programs.

CBP is interested in the nature of the inbound cargo's supply chain. This means that US importers must have documented and verifiable procedures regarding who their upstream partners are, both in and outside of the United States. Obviously, this includes all carriers; and there is also an incentive for C-TPAT-certified importers to deal with member

---

[7] Apart from being free trade partners with the United States under the *North American Free Trade Agreement*, both Canada and Mexico are, respectively, the United States' second and third largest sources of merchandise imports (by value). Together they account for about 40% of this trade flow. China, however, is the largest import source with $353 billion compared to Canada's $340 billion. Thus, C-TPAT has the potential to cover a lot, but by no means all, of inbound US cargo. The Container Security Initiative (CSI), discussed ahead, may be looked to in order to cover the rest.

[8] U.S. Customs and Border Protection (2009, p. 1). While CBP has made several thousand security validations and site visits up to 2009, it is curious that a more recent study is not available.

Table 4.2 Customs Trade Partnership Against Terrorism Best Practice Elements as Vetted by Customs and Border Protection

Best Practice Elements	Examples
Business partner security relationships	■ Random audit of supply chain partners using a third party security firm ■ Perform table-top security exercises with partners
Conveyance security	■ Use tamper-indicative security seals along hinges and seals of the doors to the container/trailer/rail car ■ Use state-of-the-art locks and bolts
Information technology security	■ Use software which allows management to delete employee laptops' hard drives from a remote location ■ Equip all computers with biometric retina scanners
Personnel security	■ Fingerprint all new employees ■ Conduct exit interviews of terminated employees with trained counselors on hand
Physical access control	■ Photo IDs must be displayed and matched to a database ■ Visitor badges should be thermal-activated to show expiration after a defined period of time ■ External doors operate on a two person, two key system
Physical security	■ Install a double-layered perimeter fence, electronically monitored with underground concrete to deter tunneling ■ Place security guard towers at all corners of the perimeter ■ Install an alarm system triggered by door contacts, heat, vibrations, and seismic activity
Procedural security	■ Use tamper-indicative tape with serial numbers to seal packages. Verify the numbers against a packing list when loading and unloading the package along the supply chain ■ Schedule deliveries in advance and have drivers provide security personnel with preassigned driver and shipment numbers for verification
Security training and threat awareness	■ Require new employees to complete a multi-module security training program ■ The training should include how to handle Internet correspondence and electronic data interchange (EDI) if applicable

Source: U.S. Customs and Border Protection, 2009. Customs-Trade Partnership Against Terrorism: Best Practices Update 2009, Washington, DC. Publication Number: 0000–0823.

carriers as well. CBP also reserves the right to conduct on-site visits to domestic and even foreign C-TPAT members' facilities.

CBP places C-TPAT applicants or members into three tiers. Tier I members have submitted their security plans and had them approved. Tier II members have their plans validated through an on-site visit by CBP officials. Tier III members are those which have, in the judgment for CBP, exceeded Tier II standards. According to CBP's own figures a Tier II member is about 3.5 times less likely to undergo cargo examination compared to nonmembers.[9] Tier III members are nine times less likely. Tier II and III members are given "front of the line" treatment when their cargo arrives at a port of entry.

Obviously, C-TPAT is not really an equal partnership since only one side can inspect and penalize the other. But, it must be said, should the US border ever be closed again, it would be C-TPAT shippers and carriers who would be allowed to resume business first. More immediate benefits are: CBP's claim of reduced number of inspections at the border (thus saving time); if a physical inspection is required the cargo goes to the "front of the line" afterward; access to a list of C-TPAT members (which is helpful in supply chain partnering); expedited land border crossing under a program called Free and Secure Trade (FAST) (discussed below); and an option to make duty payments on a monthly or bimonthly basis. Of course, going to the "front of line" could still be a rather long line of other C-TPAT cargo. Much of the benefits are subject to probability, and this must be weighed against the definite upfront costs in acquiring C-TPAT membership.

It is a tall order to extend the US zone of security all the way to the point of cargo origin, be it foreign or domestic. CBP's interest can certainly extend that far though not necessarily its jurisdiction. Thus, CBP uses a "trust but verify" approach based on the on-site foreign and domestic visits and risk assessments before the cargo reaches a US port of entry. Of course, expedited treatment at the border can free up CBP inspectors for scrutinizing more suspicious cargo on arrival since C-TPAT shipments are deemed to be low risk.

Free and Secure Trade

C-TPAT compliant motor carriers reaching the United States via Canada or Mexico are eligible to use FAST lanes at border crossings where available. Membership is open to carriers from all three countries. As to the cargo being carried, it is important that the manufacturer, shipper, driver, and US importer all be C-TPAT compliant. The intent is to make expedited clearance contingent on the international supply chain being secured via CBP site visits validating the security plans on file. Drivers

---

[9] U.S. Customs and Border Protection (2014, p. 2).

reaching a FAST-designated lane must present a special card to CBP officers in order to be cleared. The security process along the supply chain should work along these points:

1. The original manufacturer or foreign-based exporter (who bought the cargo from the manufacturer) is responsible for the cargo and placing it in a sealed container or motor carrier trailer before releasing it to the carrier. Seals will be of a high security type following ISO guidelines (i.e., ISO/PSA 17712, Freight Containers-Mechanical Seals).
2. The carrier verifies the seal number on the container/trailer with the information listed on the cargo manifest and bill of lading.
3. The carrier interfaces with CBP at a FAST-dedicated lane at a border crossing.
4. The carrier delivers the cargo to the importer noted on the bill of lading.

Out of the 122 US-Canada and 37 US-Mexico border crossing stations, each currently has 17 stations with FAST-designated lanes. The majority of the US-Canada ones are located in Michigan, New York, and Washington State; and the majority of US-Mexico FAST stations are located in California and Texas. Currently, about 78,000 motor carrier drivers in the US, Canada, and Mexico have FAST cards.

Container Security Initiative

Building on C-TPAT and the post-9/11 security mindset, the Container Security Initiative (CSI) involves agreements between the United States and partner countries which allow CBP officers to screen and inspect cargo right at the foreign ocean port of exit. CSI represents the first time that the point of supply chain security compliance was moved off of US shores to the point of foreign departure. Almost 90% of the world's manufactured cargo moves via container; and 40% of this is moved by ocean vessel. Therefore, it makes sense that the ocean vessel mode is singled out for a specific supply chain security program. Of course, as will be discussed below, screening is not the same thing as scanning, search, and inspection. In fiscal year 2015, CSI ports provided on-site CBP inspectors with about 11 million bills of lading. Based on these it was determined that about 104,000 containers be searched and examined.[10] This is, of course, only about 1%.

---

[10] Kulisch (2014, p. 40).

> **Table 4.3** Container Security Initiative Ports
>
> **Canada**: Vancouver, Montreal, and Halifax
> **Europe**: Algeciras, Antwerp, Barcelona, Bremerhaven, Felixstowe, Genoa, Gioia Tauro, Gothenburg, Hamburg, La Spezia, Le Havre, Lisbon, Liverpool, Livorno, Marseille, Naples, Piraeus, Rotterdam, Thamesport, Tilbury, Southampton, Valencia, and Zeebrugge
> **South/Central America**: Balboa, Buenos Aires, Caucedo, Cartagena, Colon, Freeport, Kingston, Manzanillo, Puerto Cortes, and Santos
> **Asia**: Chi-Lung, Colombo, Hong Kong, Kaohsiung, Kobe, Laem Chabang, Nagoya, Port Klang, Pusan, Singapore, Shanghai, Shenzhen, Tanjung Pelepas, Tokyo, Yantian, and Yokohama
> **Africa**: Alexandria and Durban
> **Middle East**: Ashdod, Dubai, Haifa, Qasim, and Salalah

To date CSI operates in 59 ocean ports around the world as listed in Table 4.3; and this covers around 80% of the cargo destined for the United States. After CBP has done its screening at the port, the cargo manifests and the other ISF materials are received electronically in the United States via ACE at CBP's National Targeting Center (NTC) in advance of cargo arrival. This is supposed to give CBP time to further screen data in order to determine which containers should be physically inspected on arrival. If further screening is deemed necessary while the cargo is en route, the containers can be put through large-scale X-ray machines, gamma ray machines, and radiation detection devices at the port of entry. For those ports not yet in the CSI program the advance manifests are nonetheless reviewed at the NTC. Apart from identifying shipments potentially used in terrorism the NTC searches for evidence of smuggling, human trafficking, counterfeiting, and money laundering over all modes of international transportation.

Eleven million containers arrive at US ocean vessel ports. This represents about 10% of the 108 million containers arriving at ocean ports around the world. Given these numbers CBP is reviewing data on around 30,000 containers per day. The majority of US imports arrive at ocean ports with the shares being 52% by value and 75% by weight.[11] CBP also tries to tailor CSI to any special features of the port in question. These differences could involve the particular origin or content of

---

[11] Pocket Guide to Transportation (2013).

the majority of the cargo. It could involve how the vessels are unloaded [e.g., lift-on, lift-off (LO-LO) or roll-on, roll-off (RO-RO)]; or it could involve intermodal activity (e.g., transferring containers to motor carriers or to rail). Each of the operational considerations affects the time for loading and unloading of containers.

Security and Accountability for Every Port Act

The Security and Accountability for Every Port Act (2006),[12] also referred to by the acronym SAFE Port Act, set out the statutory framework for C-TPAT. Another provision of the act requires that containers entering 22 of the busiest ports be scanned for radiation. This could be by physical inspection or by X-ray and gamma ray detection machines. The act also requires a 100% scanning rate at all ports of entry; but such compliance is a long way off since many ports do not have the scanning equipment in place or, in some cases, the necessary physical space to screen incoming containers. Alternatively, using CSI ports as a means to scan at the port of exit does not help much either since many of these foreign ports have the same challenges regarding physical space. Furthermore, relying too much on foreign ports presents another problem. If an item is scanned at a CSI port but is transshipped through another foreign port (in CSI or not) on its way to a US port of entry, DHS is rightly concerned about the possibility of security breaches during that layover. If the transhipment port is not in CSI, then vulnerability has risen; but even if it were a CSI port it is challenging to stage scanning equipment in a timely and nonintrusive manner in the transhipment area. Recall, that CSI is designed to deal with US-bound containers arriving from inland of the country in question before being released to the container vessel. With transshipping, on the other hand, the container is being offloaded from the vessel and moved into a nearby holding area among thousands of other containers.

In order to comply with the Congress, DHS originally wanted 100% compliance to be achieved in 2012; however, it asked for, and received, approval from the Congress to extend the deadline to 2014. Another request for delay was made in May 2016. It is expected

---

[12] The act is set out as Public Law 109–347—OCT. 13, 2006. It may be found at: https://www.congress.gov/109/plaws/publ347/PLAW-109publ347.pdf.

that the Congress will extend the deadline to May 2018.[13] Currently, about 5% of all inbound containers are scanned; therefore, compliance will be a huge logistical challenge. Also, while 22 ports was the original intent of SAFE, US imports arrive from about 750 ports worldwide.

The act set out some important definitions and distinctions concerning how CBP will deal with a container:

- **Scan**: utilizing nonintrusive imaging equipment, radiation detection equipment, or both, to capture data, including images of a container.
- **Inspection**: the comprehensive process used by CBP to assess goods entering the United States to appraise them for duty purposes, to detect the presence of restricted or prohibited items, and to ensure compliance with all applicable laws. The process may include screening, conducting an examination, or conducting a search. These are:
  - **Screening**: a visual or automated review of information about goods, including cargo manifest or entry documentation accompanying a shipment being imported into the United States, to determine the presence of mis-declared, restricted, or prohibited items and assess the level of threat posed by such cargo.
  - **Examination**: an inspection of cargo to detect the presence of mis-declared, restricted items, or prohibited items that utilize nonintrusive imaging and detection technology.
  - **Search**: an intrusive examination in which a container is opened and its contents are removed and visually inspected for the presence of mis-declared, restricted, or prohibited items.

In other words, scanning involves the capture of data and images involving the container and its contents without actually opening the container. Inspection involves assessing the container for customs compliance. Inspection may be conducted in three ways. First, screening involves reviewing just the documentation accompanying the container. Second, examination is a scan used to comply with the goals of

---

[13] On the issue of delays see Gallagher (2016, p. 76). On the issue of the lack of technology available to effectively provide 100% compliance, see Hutchins (2016b, p. 26).

an inspection—a nonintrusive examination of the container. Third, a search is, in effect, a physical inspection of the container's contents.

Advance Cargo Manifests

One of the first things to change in post-9/11 transportation was the requirement to provide information accurately, electronically, and in advance of transporting international containerized cargo to the point of domestic entry. All modes of transport (except for pipelines) were affected. Gone were the days of shippers and carriers listing the shipment as "Freight All Kinds" or "Said to Contain" on cargo manifests.

Cargo manifests should not be confused with either bills of lading or waybills. A bill of lading is a legal contract between a shipper (consignor) and carrier noting: title of the cargo; how it is to be transported and turned over; and how the service is to be priced. It is usually prepared by the shipper and signed by the carrier; thus, making it a type of receipt, too. Once the cargo is delivered and the bill of lading is handed to the recipient (consignee) the title is transferred (assuming the shipment was consigned "to order"). Waybills are similar to bills of lading, and can act as a receipt, but they cannot be consigned "to order" meaning that they cannot be used on their own to transfer title except to a prespecified consignee. Waybills, therefore, are used when the consignee does not require legal documents either because: (1) the consignee has paid the consignor in advance; or (2) the consignor and consignee have a strong business relationship. Finally, a manifest simply lists the cargo (item by item) in the conveyance and other items related to the voyage—it is basically travel data covered in the bill of lading and the waybill.

Each mode has its own set of rules to be followed. The information is to be provided to CBP in advance and in electronic format using ACE. Key elements to be provided on the import cargo manifest are:

1. Foreign port of departure
2. Standard Carrier Alpha Code (SCAC)
3. Voyage number of the vessel
4. Date and place of scheduled arrival in the first US port
5. Number and quantity of packages based on the bill of lading (i.e., the contract between the shipper and the carrier or forwarder)
6. First port of receipt by the carrier

7. Cargo description: detailed shipper description or a 6-digit harmonized tariff schedule number; and the cargo's weight
8. Shipper(s) name(s) and address(es) or ID numbers assigned by CBP
9. Consignee(s) names(s) and address(es) or ID numbers assigned by CBP (i.e., these are the party or parties that will accept delivery of the cargo in the United States. The shipper is the consignor)
10. Name of the vessel, national flag, and vessel number
11. Names of foreign ports visited beyond point (6) above
12. International hazardous goods code if applicable to the cargo
13. Container number
14. Numbers on all seals affixed to the container

Table 4.4 lays out the set of advance electronic manifest rules required under Section 343 of the Trade Act of 2002. One thing to notice is that the United States Postal Service (USPS), as a government entity, is exempt from providing information to CBP while a private sector commercial carrier is required. When the shipments are inbound, the onus is on the commercial carrier to provide the electronic manifest to

Table 4.4 Advance Electronic Manifest Rules by Mode

Mode	Final Rules (Departure and Arrival Bases) (Effective Oct. 1, 2003)		Responsible Party to Send Manifest to Customs and Border Protection	
	Inbound	Outbound	Inbound	Outbound
Air (for cargo ≥1 lb)	4 h (Before arrival) or "wheels up" if flight is < 4 h	2 h (Before departure)	Air carrier	Exporter
Rail	2 h (Before border arrival to enter)	2 h (Before border arrival to exit)	Rail carrier	Exporter
Motor carrier	30 min (Before border arrival to enter; if FAST compliant)	30 min (Before border arrival to exit; if FAST compliant)	Motor carrier	Exporter
	1 h (Before border arrival to enter)	1 h (Before border arrival to exit)	Motor carrier	Exporter
USPS	*exempt*	*exempt*	*N/A*	*N/A*
Ocean vessel	24 h (Prior to lading at foreign port)	24 h (Prior to departure from US port where cargo was laden)	Ocean carrier	Exporter

CBP. This also means that the carrier has to collect accurate information from its shipper customer or any intermediaries acting on behalf of the shipper. When the shipments are outbound the onus is on the exporter and not the carrier.

Air cargo rules for inbound cargo differentiate between flights that are longer than or shorter than 4 h. Rules for motor carrier containerized freight differentiate between whether or not the parties are members of FAST. Inbound ocean vessel containerized cargo has the strictest of all the rules—requiring submission 24 h before the container has even been placed on the vessel at the foreign port. All of the other inbound rules are arrival based. It is odd that the ocean vessel rule, being departure based, treats cargo departing from, say, the Port of Vancouver in Canada destined for the Port of Tacoma with the same time constraint as cargo from the Port of Hong Kong. While the trip from Hong Kong would be much longer, giving more time to process the manifest, CBP's rationale seems to focus on screening cargo at the origin and not the destination.

Bulk shipments (e.g., coal, grain, oil, lumber, steel beams, etc.) and palletized shipments are exempt from the advance manifest rule which applies to containers. CBP defines bulk cargo as homogenous cargo that is stowed loosely in the hold of the vessel and is not enclosed in a container. However, electronic manifests must reach CBP 24 h before the ship reaches a US port. Even empty containers being hauled as part of repositioning plans may require notice to CBP 48 h prior to arrival.

DISASTER RELIEF PROGRAMS

The Basics

Short of fighting wars a government's role in society is never tested as hard as when it is responsible for disaster relief. Since most disasters are unpredictable, because nature itself is unpredictable, so is the demand for aid with respect to its composition, the volumes of what is needed, and the location it needs to be sent. Responses also need to be immediate. In commercial supply chain management and logistics management an untimely response can lead to such problems as transportation congestion and lost sales. However, when facing a disaster an untimely response leads to loss of life.

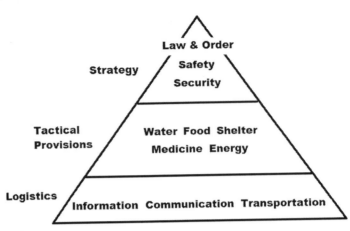

FIGURE 4.1

Support structure for disaster relief.

In a disaster environment the government must maintain law and order and keep those impacted safe and secure. This should be the government's strategic goal. In order to achieve this those impacted need to be provided with five crucial items: water, food, shelter, medicine, and energy (power). Without any one of these it is hard to imagine law and order being restored in a peaceful way. Therefore, provision of these five items should be part of the government's tactics. Of course, logistics is key to successfully providing any item over some distance and area; and it is especially so if the area has been hit with a disaster of some kind.[14] Logistics represents a foundation for the strategy and tactics the government wishes to undertake. Indeed, the logistics infrastructure itself may be impacted severely by the disaster. Nonetheless, the logistics foundation is comprised of three items: information, communication, and transportation. Without any of these items, the coordination and delivery of relief is not possible. Therefore, the government needs to be concerned with the logistics which feed into its humanitarian supply chain. Fig. 4.1 gives a visual representation of this structure.

The government's relief plan needs to replace the impacted commercial flow of goods, services, and information until markets can be restored. In order to get markets working again, people need to have their normal lives restored. Families have to be brought back together; workers

---

[14] Humanitarian logistics is an applied field. Good examples of this research include: Atkinson and Sapat (2012); Matopoulos et al. (2014), and Trestrail et al. (2009).

need to return to their jobs; and children need to return to school. Such an endeavor requires large-scale planning and coordination with many organizations. This is very much an application of supply chain planning and management.

Two Approaches to Planning

If a large-scale disaster takes place, all eyes turn to the government for assistance, relief, and reconstruction. One approach to meeting the government's goal is to be proactive and plan in advance. The other is to be reactive with a looser plan which adapts to the specific needs of the situation. From a supply chain and logistics point of view it is possible for either of the two approaches to be successful. The proactive approach requires the right private sector partners and vendors to be in place and able to deal with the disaster in question. They have to be able to broadly deal with the tactical and logistical challenges noted in Fig. 4.1. The reactive approach, on the other hand, requires the government to know quickly which private sector partners and vendors to tap for the relief effort. More specialized firms may be recruited in this case; but time will be of great essence.

There is a certain tension between these two approaches from a logistics point of view. A proactive approach would necessitate staging elements for relief efforts near anticipated impact areas across the country. This is similar to push logistics and buy-to-stock tactics. The reactive approach would hold these elements in reserve and send them out to the impact area once it has been identified. This is similar to pull logistics and buy-to-delivery tactics. John W. Madden, former director of the Alaska Division of Homeland Security and Emergency Management (DHSEM), sees this as a logistical choice between "just in case" versus "just in time." However, under his leadership, Alaska adopted a hybrid approach.[15] Why? Since some parts of Alaska do not even have the necessary logistics infrastructure to move to an impacted area (especially in winter months), it was deemed prudent to stage water, durable goods, and fuel at several rural communities. Other less durable relief items would be centrally held and released as needed. The hope would be that the staged items would provide enough immediate relief to hold

---

[15] See Madden (2014). Alaska is a state which has many vulnerabilities. Large distances over wilderness and mountain ranges are combined with sparse rural infrastructure. This is combined with susceptibility to volcanoes, earthquakes, floods, and, depending on the season, bitter cold and wildfires.

out for the inbound items. In 2005, Alaska DHSEM formed a PPP to help anticipate needs and assist in relief efforts. The Alaska Partnership for Infrastructure Protection includes the private owners of key infrastructure in Alaska to work with the Alaska government. Membership is voluntary and the incentive to join is the feeling that the investment of time and energy in helping to plan is exceeded by the mitigation of risk the private sector partners feel they are gaining.

Federal Emergency Management Agency

DHS was formed in the wake of 9/11 in order to bring various government, trade, and security organizations under one roof. The Federal Emergency Management Agency (FEMA) was formed in 1979 with a similar intent— to bring several emergency preparedness and response organizations under one roof. Communicating with the federal government became so cumbersome and disjointed that the National Governors Association petitioned the president to consolidate the necessary organizations. These included:

- Defense Civil Preparedness Agency (formerly under the Department of Defense)
- Federal Broadcast System (formerly in the Office of the President)
- Federal Disaster Assistance Administration [formerly under the Department Housing and Urban Development (HUD)]
- Federal Insurance Administration (formerly under HUD)
- Federal Preparedness Agency (formerly under the General Services Administration)
- National Fire Prevention Control Administration (formerly under the Department of Commerce)

All natural and man-made disasters would fall under its jurisdiction. Originally, the FEMA director would report directly to the president; however, in 2003 FEMA would be absorbed into DHS and the director now reports to the DHS Secretary. This changed FEMA's primary mission to national security and terrorism from its original "all-hazards" approach.[16]

The biggest challenge FEMA faces is coordination with state and local authorities. This is very much akin to a governmental supply chain

---

[16] During the Reagan Administration FEMA's mission was changed to a primary focus on nuclear warfare. For a fuller discussion see Bullock et al. (2013, pp. 615–616).

problem—coordinating with one's vendors (i.e., the private sector sources of emergency supplies) and one's customers (i.e., the state/local governments and the civilians impacted). The most prominent examples of failure to coordinate include the responses to Hurricane Andrew in Florida and Louisiana (1992) and Hurricane Katrina in Louisiana (2005). The problem is one of balancing the abilities of the federal government's access to vast supplies and the local government's more intimate knowledge of the needs of the local population. It also does not help that the federal government is burdened with many more regulatory hoops to jump through than local governments. For example, the Disaster Relief and Emergency Assistance Act (or Stafford Act) of 1988 gives FEMA the power to directly allocate funds to local governments and firms. In reality, however, the federal bureaucracy has made it difficult to put this into practice.[17]

A part of the overall choice for proactive or reactive planning is FEMA's choice of private vendors to be used in relief efforts. It is certainly proactive to have vendors in place and on contract. Contracts could be short-term (such as for bottles of water) or long-term (when reconstruction is necessary). Vendors could be chosen based on competitive bidding in order to keep prices low and allow FEMA to demonstrate fiscal responsibility. Pre-disaster planning is more of a buyer's market; and unlike other agencies FEMA is not required to solicit competitive bids. The challenge, of course, is to anticipate needs correctly. FEMA, however, has tended to use the reactive approach. With prompt response taking precedent over other goals there is the chance for incorrect, insufficient, and overpriced aid being gathered. In an emergency the vendors are in a seller's market. "Use of non-competitive procurement approaches increased at the federal level by 115% from 2000 to 2005, and $8.7 billion was awarded through no-bid contracts for emergency procurement."[18] The value of a hybrid approach becomes evident when considering local vendors. In a proactive approach it is hard to know which vendors may be able to provide service in the vicinity or region of a disaster. In a reactive approach, on the other hand, when a disaster strikes it is even harder to reach out to the available vendors because communication technology may be compromised and, furthermore, there was no coordinating

---

[17] Atkinson and Sapat (2012, p. 360).
[18] Atkinson and Sapat (2012, p. 362).

plan put in place for the vendors to respond to. It is also noteworthy that: "The Army Corps of Engineers reported receiving over 6300 phone calls within two weeks after (Hurricane) Katrina landed, many from local and regional contractors who have complained that their calls (to assist in the relief efforts) were ignored."[19] The hybrid approach would substitute preapproved vendors with local/regional ones if they can demonstrate the ability to provide the necessary services in a timely fashion. Of course, employing vendors as close to the disaster as possible has two desirable effects: (1) it helps to lower transport costs; and (2) locals are helping locals which is necessary in sustaining both the economy and the social fabric.

DO THE PROGRAMS WORK?

It is difficult to be definitive about the efficacy of crime prevention and it is especially so with respect to terrorism and disaster relief. The problem is the *post hoc, ergo propter hoc* fallacy. This can be understood using the old joke where someone is seen with a banana in his ear. When asked why he has a banana in his ear he says it is in order to keep the alligators away. When he is told there are no alligators around he exclaims: "See, it works!" So, with this in mind, when we see more and more regulations along with more and more money going into supply chain security programs, can we exclaim: "See, they work!"? It is true that there has not been a 9/11 style terrorist incident in the United States in over 15 years now. The government can, of course, fend off the charge of *post hoc, ergo propter hoc* by stating that the successes are classified information. Therefore, success is a matter of public trust— trust in the policing power of the government. Of course, the government also wishes to partner with the private sector. While the government may be dependent on the private sector for success, the partnership is asymmetric in terms of information sharing and power. The symmetry, however, is reversed in a reactive disaster relief approach. Relief will be prompt but can be hurried and uncoordinated, and with the vendors themselves maintaining control of price and quality of the products. The alternative is to plan ahead and coordinate with state and local governments in the process. The

---

[19] Cray (2005, p. 20).

complexity is higher but breaking down silos is at the heart of effective supply chain management.

What is important is to recognize the information asymmetry problem and incentivize the behavior that the parties desire from each other. Consider C-TPAT since it is an optional program. It costs time and money for shippers and carriers to comply with CBP's expectations for membership. Therefore, only suppliers and carriers which feel they have very good supply chain security plans and procedures would take the time to design and submit their plans to CBP. For its part, CBP promises expedited treatment, access to FAST lanes, etc. C-TPAT acts as a screening mechanism in order to separate those needing more scrutiny from those who need less. Of course, a carrier which is hauling multiple shippers' loads on a single truck [i.e., a less-than-truckload (LTL) carrier] needs all shippers to be C-TPAT members in order to use the FAST lane at the border. Thus, the shipper-carrier market for inbound US transport is divided into two classes. This is an expected result since CBP faces the adverse selection problem. Of course, a related problem is moral hazard. How can CBP know that the security plans will be adhered to now that the shippers and carriers are given the benefits up front? Certainly, periodic inspections will help. But this is CBP in the role of policeman. Another way is to bring the shippers and carriers into the conversation over how to secure and regulate supply chains. Since shippers and carriers desire security, too, it makes sense for CBP to demonstrate partnership. The question is can PPPs remain stable when the public entity can change hats between partner and policeman? Is CBP willing to bind itself in a sort of contract with shippers and carriers in the way shippers and carriers bind themselves?

ACE is supposed to create a "single window" linking the private sector to the government (i.e., to all 48 individual agencies). This will be a major application of ERP theory. The technology behind the portal will be of little value unless all parties can and are willing to cooperate and share information. SAFE requires 100% of all incoming containers from 22 ports to be scanned. The date of compliance is being pushed ahead. Obviously, technology has not caught up with the government's expectations. It is too soon to tell whether or not these programs will work to the benefit of secure international supply chains leading into the United States.

ISSUES AND PROBLEMS GOING FORWARD

Technology only works if the people using it share the same goals and are willing to trust and cooperate. At the same time, technology used to scan and screen cargo creates an automated process which may come at the expense of human detection and experience. The process may be faster but is it making the supply chain more secure? Paperwork, albeit in a scanned electronic format, can be read and analyzed by data mining algorithms or by human beings. Chapter 5 looks at the data gathering process and the threats to it. Data flow is an important part of the risk mitigation process. Whether an item is in storage or in transport, data is being gathered. The supply chain security and disaster relief programs are, or have many attributes of, PPPs. Chapter 6 looks at these in more detail in order to better gauge the level of cooperation involved.

Bibliography

Atkinson, C.l., Sapat, A.K., 2012. After Katrina: comparisons of post-disaster public procurement approaches and outcomes in the New Orleans area. Journal of Public Procurement 12 (3), 356–385.

Bullock, J.A., Haddow, G.D., Coppola, D.P., 2013. Introduction to Homeland Security, fourth ed. Butterworth-Heinemann, Waltham, MA.

Cray, C., 2005. Disaster profiteering: the flood of crony contracting following Hurricane Katrina. Multinational Monitor 26 (9), 19–24.

Edmonson, R.G., March 25, 2009. Known shipper program not working. Journal of Commerce 15.

Gallagher, J., May 30, 2016. Help wanted: homeland security seeks private company help to meet 100 percent container-scanning deadline. Journal of Commerce 76.

Hutchins, R., January 11, 2016. ACE nears launch date. Journal of Commerce 72.

Hutchins, R., July 25, 2016. 100 percent fired up: the debate over scanning of all U.S.-Bound containers flares up again in Congress. Journal of Commerce 26.

Kulisch, E., July 19, 2014. CSI's evolution. Journal of Commerce 40.

Madden, J.W., July 2014. Critical infrastructure in Alaska: a system of competitive collaboration. The CIP Report 13–14.

Matopoulos, A., Kovacs, G., Hayes, O., 2014. Local resources and procurement practices in humanitarian supply chains: an empirical examination of large-scale house reconstruction projects. Decision Sciences 45 (4), 621–646.

Pocket Guide to Transportation, 2013. Bureau of Transportation Statistics (BTS), Washington, DC. Online. Tables 4-6 and 4-7. http://www.rita.dot.gov/bts/publications/pocket_guide_to_transportation/2013.

Security and Accountability for Every Port Act, 2006. Public Law 109–347—OCT. 13, 2006. www.congress.gov/109/plaws/publ347/PLAW-109publ347.pdf.

Trestrail, J., Paul, J., Maloni, M., 2009. Improving bid pricing for humanitarian logistics. International Journal of Physical Distribution & Logistics Management 39 (5), 428–441.

U.S. Customs and Border Protection, 2016. ACEopedia. Washington, DC. Publication Number: 0293–1015.

U.S. Customs and Border Protection, 2014. C-TPAT Program Benefits: Reference Guide. Washington, DC. Publication Number: 0192–0114.

U.S. Customs and Border Protection, 2009. Customs-Trade Partnership Against Terrorism: Best Practices Update 2009. Washington, DC. Publication Number: 0000–0823.

CHAPTER 5

The Role of Information Technology

CONTENTS

FROM DATA TO WISDOM

A lot of data is being collected today. Some of it may be used meaningfully and some of it may be simply ignored or discarded if organizations do not have the time or training to analyze it. Before examining the technology used to collect and process data, it is worthwhile to consider how users of data and technology should

Global Supply Chain Security and Management. http://dx.doi.org/10.1016/B978-0-12-800748-8.00005-4

proceed along a knowledge chain intended to add value each step of the way. The following stepwise hierarchy is suggested:[1]

1. **Data** involves symbols. It is raw and has no meaning on its own. No pattern recognition has been performed. Everything is still noise.
2. **Information** is data that has been processed by the user in order to be useful in decision making. Context or connections made to other data has provided meaning. Information answers questions like "who," "what," "when," and "where." Having information is a necessary condition for pattern recognition.
3. **Knowledge** is information processed by the user to answer the question of "how." A system is created which collects related information and adds to that knowledge base. Having knowledge is a sufficient condition for pattern recognition.
4. **Understanding** means using knowledge to theorize and explain "why." Not only is there a system but it can be evaluated against different systems in order to either invalidate one of them, determine their independence, or integrate them into a richer knowledge base. This involves analysis and learning. Having an understanding allows for pattern recognition to be extended in order to determine cause and effect.
5. **Wisdom** is understanding which has been evaluated, substantiated, or proved. This process builds in ethics and culture in order to achieve a less time-specific understanding of the human condition. Having wisdom takes the search for cause and effect and extends from a local context to a global one.

Points 1–4 relate to the past and indicate what is known to the decision maker. Wisdom relates to the future, because it involves validation, propositions, or forecasts. In a supply chain security context, for example, these five steps could proceed in this way:

1. **Data**: items in transit have been stolen (fact/observation).
2. **Information**: items in transit have been stolen on routes which include a vehicle lay over (possible cause and effect).
3. **Knowledge**: items in transit have been stolen on routes when the vehicle is laid over at night or in a high crime area. (A connecting pattern emerges which strengthens the case for cause and effect.)

---

[1] This section draws on the work of Ackoff (1989).

4. **Understanding**: vehicles laid over during the day or in secure areas do not experience stolen items. (Cause and effect becomes clearer when one system of transport is compared to an alternative.)
5. **Wisdom**: crime is a fact of life and it thrives on opportunity. Night and high crime areas both incentivize criminal behavior. A vehicle containing valuable items laid over in the open provides an opportunity for a criminal to ply his skill. Incentive and opportunity are key pieces to crime's cause and effect (learning/experience).

These five steps can be used as means to evaluate the efficiency of the data collection process, the way the technology stores the data, and the way business and government use it in order to try to turn data into wisdom. Of course, this five-step process is not intended to be effortless. The data collected must be accurate, timely, and in a quantity sufficient to make a meaningful impression on the user. At least two data points are needed in order to establish a pattern; however, most decision makers are not comfortable with declaring a pattern until they see many more data points. The emergence of Big Data is a solution to a decision maker's demand for evermore data. Of course, moving from knowledge to understanding and wisdom involves a solid appreciation for the process of cause and effect. Recall the *post hoc, ergo propter hoc* fallacy noted in Chapter 4. Correlation and sequencing does not necessarily indicate causation. Theorizing to explain the "how" and the "why" involves deep learning and a lot experience with the issue being investigated. Theory can be developed deductively (i.e., testing a theory against the collected facts) or inductively (i.e., using the collected facts to let the theory emerge).[2] In previous decades data and computing limitations meant that a lot of theorizing was deductive; but technology is beginning to tip the scale toward inductive theorizing. Data mining now applies a variety of statistical techniques to try and uncover patterns in very large datasets. But what has really increased the value of data mining is combining it with machine learning (or artificial intelligence), whereby the computer algorithms are automatically adjusted in order to try to "learn" from the continuous stream of incoming data. Whichever method the decision maker chooses, he must always be a learner and adapting whether it be on an instinctual basis or an experiential basis.

---

[2] For a discussion of deduction and induction in a transportation context, see Prentice and Prokop (2016, pp. xviii–xxii).

CYBERSECURITY

The Information Supply Chain

FedEx founder Fred Smith is famous for noting that information about a package is just as important as the package itself. This means that carriers ought to consider themselves as being in the information business as much as they are in the transportation business. Of course, modern supply chain management notes the coexistence of these two services.

Information availability in today's information age coupled with the increased speed to send/receive are part of the reason for modern supply chain management. Searching for vendors or customers is simply a matter of finding a webpage or web directory. Communicating via email is about the same process and cost, no matter if the point of contact is across the street or across the globe. These new lines of communication make it easier for all parties in a supply chain to be more particular about their partners, and more flexible in building relationships; after all, it is easy to go online and find another vendor, another place to set up production, or another locale to market goods. More sensitive information has been placed online than ever before. Business transactions are becoming more and more paperless.

Supply chain management involves the movement of tangibles (e.g., natural resources, sub-assemblies, and final products). However, such activity is preceded by movement of information (e.g., buy/sell requests, contracting negotiations, and documents). Information may be exchanged post transaction as well. Today, a growing share of the trade in information is taking place on the Internet instead of on paper or on the telephone. This increases the speed of the transaction; but it also opens up these private activities to computer hacking, viruses, and theft. Therefore, it makes sense to extend supply chain security from the world of brick-and-mortar to the world of cyberspace. Of course, this is a challenging endeavor. As brick-and-mortar supply chains take on more partners/connections, complexity grows. In the world of the Internet, a computer or wireless device conceivably has millions of connections. Two often contradictory tasks must coexist: the need to share information and the need to protect it.

The bundling of information and tangibles in order for the supply chain to function fits in with the literature on the resource-based view

(RBV) of a firm's competitive advantage. RBV contends that bundles are necessary to create "capabilities," and it is these which are the source of competitive advantage.[3] RBV studies in supply chain management have concentrated on how to improve the robustness of the supply chain through information sharing.[4] In fact, some proponents of RBV refer to this interorganizational view as the knowledge-based view.[5] In this context, a supply chain security plan, well understood and deployed both upstream and downstream among the interconnected organizations, is another specific "capability."

Consider the Automated Commercial Environment (ACE) discussed in Chapter 4. Customs and Border Protection (CBP) issued many delays as technology was developed and industry was given time to adjust to the new requirements for this projected "single window" of trade. In 2012, CBP decided to roll-out the ACE software in small doses instead of in a full form and having to incorporate the requirements of the 48 different government agencies, which oversee international trade. The 2012–17 period has, in effect, been akin to an open source program (deploying so-called "agile" software development) instead of some proprietary software developed in isolation. Given the complexity of trade and how it is regulated, this was a wise decision. As all private sector players try out the system, it gives them a chance to compare ACE to the legacy systems and give feedback to CBP's programmers to help improve the system.[6]

The Computer Fraud and Abuse Act (1986) was the first comprehensive attempt, regarding computers used in interstate commerce or in the federal government itself, to prevent cybercrime. Offenses included unauthorized access to: state secrets, financial information, medical information, and computer passwords. The Federal Information Security Management Act (2002) and its updated version, the Federal Information Security Modernization Act (2014) gives Department of

---

[3] For a discussion of RBV and capabilities, see Barney (1991).

[4] For a discussion of supply chain visibility in the RBV context, see Brandon-Jones et al. (2014).

[5] See Mellat-Parast and Spillan (2014).

[6] CBP is by no means the only customs organization experiencing difficulty with its government partners, shippers, and carriers in trying to develop a single window for electronic communication. For a review of the issues in other countries see Urciuoli et al. (2013).

Homeland Security (DHS) oversight in how various government agencies implement cybersecurity plans.

Cyberspace and Security

Cyberspace encompasses an interdependent network of information technology (IT) infrastructures. This includes the Internet, telecommunications networks, and interconnected computer networks via cable or wireless. People and businesses want access to it because of its convenience in finding information and communicating across the world. Of course, if one can easily access cyberspace a criminal can just as easily target the user. As noted in Chapter 3 many people feel a quasi-personal connection to the cyber-world through sharing information both personal and transactional. This ease opens up vulnerabilities.

Cybersecurity involves preventing infiltration, damage to, or unauthorized use of electronic information and communication systems. It also serves to insure the confidentiality, integrity, and availability of the information. DHS also includes in its own mandate the protection and restoration, when needed, of information networks, wireless, satellite, public safety answering points, and 911 communications systems.

Global connectivity of computers along an information superhighway means a lot of information can be gathered and shared. Also, a lot of things can occur as the Internet is searched; and these can put security at risk. Web-based software such as Java allows for interactive web pages which increase functionality. However, these very features can be used by web-based attackers to install spyware on the computer which reached out to the website. Active content and cookies (which identify the user to the owner of the webpage being visited) can be disabled through tighter privacy and security settings on the computer. Of course, the trade-off is a loss in some Internet functionality.

Cybersecurity, however, is not easy. This is because cyberspace was designed first and foremost to promote connectivity. As the commercial Internet developed out of its affiliation with the US military, security did not seem to be an issue of concern for the early designers. All attempts to secure the Internet, from antivirus software, patches, updates, firewalls, etc., have been reactive as opposed to proactive. Another reason is the competitive nature of the software industry.

There is an incentive to issue software quickly in order to beat rivals. If any flaws are discovered they can be handled by software updates. This leads to vulnerability because their user customers may not run updates fast enough or often enough. Of course, network connectivity means that a hack into one user can quickly spread to others. In a supply chain context, cybersecurity is a public good and the market itself has an incentive to under provide it. This is a market failure which is hard to correct.[7]

Finally, the life blood of IT and telecommunications is the electricity which powers it. Electricity is generated via burning fossil fuels, nuclear fission, or hydro activity. Once generated, it is transmitted and distributed by electricity utility companies (which many be public or private) via power lines. This electricity supply chain, or power grid, is an important foundation to the information flow along any supply chain. But the power grid itself is subject to cyberattack. Therefore, the securing of commercial supply chains must also take into account all underlying infrastructure.

Cybercrime

While information is intangible, it is also valuable. It is just as susceptible to theft as is physical inventory. The FBI noted that in 2014 alone 519 million financial records have been stolen from banks and 110 million Americans have had their identities open to theft. The US government itself reported 67,168 incidents within its own agencies in 2014 as well.[8] The U.S. Government Accountability Office (GAO) divides cybercriminals into the categories as shown in Table 5.1.

These cybercriminals use a variety of techniques to gain access to computer networks. The more prominent types are noted in Table 5.2.

---

[7] Basic economics recommends that government correct market failures. So far, the government has shied away from taxes and subsidies to create the necessary incentives. Instead, it has chosen regulation. The US federal government maintains a security breach law covering interstate commerce. Most of the states (except for Alabama, New Mexico, and South Dakota) maintain similar laws. For a comparison, see Steptoe and Johnson LLP (2016). These laws require disclosure of breaches to all those ill-affected. In this way, a firm's customer base or vendors will be made aware and react accordingly (i.e., either contractually or through litigation).

[8] See Kelly (2014) and U.S. Government Accountability Office (2015, p. 7).

Table 5.1 Cybercriminal Categories

Category	Description
Bot-network operators	They use a network (colloquially a bot-net) of compromised, remotely controlled systems to coordinate attacks and to distribute phishing schemes, spam, and malware attacks. The services of these networks are sometimes made available on underground markets (e.g., purchasing a denial-of-service attack or services to relay spam or phishing attacks).
Criminal groups	They seek to attack systems for monetary gain. Specifically, organized criminal groups use cyber exploits to commit identity theft, online fraud, and computer extortion. International corporate spies and criminal organizations also pose a threat to the United States through their ability to conduct industrial espionage and large-scale monetary theft and to hire or develop hacker talent.
Hackers and hacktivists	They break into networks for the challenge, revenge, stalking, or monetary gain, among other reasons. Hacktivists are ideologically motivated actors who use cyber exploits to further their political goals. Though gaining unauthorized access once required a fair amount of skill or computer knowledge, hackers can now download attack scripts and protocols from the Internet and launch them against victims' sites. Thus, while attack tools have become more sophisticated, they have also become easier to use. According to the Central Intelligence Agency, the large majority of hackers do not have the requisite expertise to threaten difficult targets such as critical US networks. Nevertheless, the worldwide population of hackers poses a relatively high threat of an isolated or brief disruption causing serious damage.
Insiders	Disgruntled insiders may not need a great deal of knowledge about computer intrusions because their position within the organization often allows them to gain unrestricted access and cause damage to the targeted system or to steal system data. The insider threat includes contractors hired by the organization, as well as careless or poorly trained employees who may inadvertently introduce malware into systems.
Sovereign nations	Nations use cyber tools as part of their information gathering and espionage activities. In addition, several nations are aggressively working to develop information warfare doctrines, programs, and capabilities. Such capabilities enable a single entity to potentially have a significant and serious impact by disrupting the supply, communications, and economic infrastructures that support military power—impacts that could affect the daily lives of citizens across the country.
Terrorists	They seek to destroy, incapacitate, or exploit critical infrastructures in order to threaten national security, cause mass casualties, weaken the economy, and damage public morale and confidence. Terrorists may use phishing schemes or spyware/malware in order to generate funds through extortion or gather sensitive information.

Source: U.S. Government Accountability Office, 2015. Cyber Threats and Data Breaches Illustrate Need for Stronger Controls across Federal Agencies. U.S. Government Accountability Office, Washington, DC, p. 4.

Table 5.2 Cybercrime Techniques

Category	Description
Cross-site scripting	An attack that uses third-party web-based resources to run script within the victim's web browser or scriptable application. This occurs when a browser visits a malicious website or clicks a malicious link. The most dangerous consequences occur when this method is used to exploit additional vulnerabilities that may permit an attacker to steal cookies (data exchanged between a web server and a browser), log key strokes, capture screen shots, discover and collect network information, and remotely access and control the victim's machine.
Denial-of-service or distributed denial-of-service	An attack that prevents or impairs the authorized use of networks, systems, or applications by exhausting resources. A distributed denial-of-service attack is a variant of the denial-of-service attack that uses numerous hosts to perform the attack.
Malware	Known as malicious code and malicious software. It refers to a program that is inserted into a system, usually covertly, with the intent of compromising the confidentiality, integrity, or availability of the victim's data, applications, operating system, or otherwise annoying or disrupting the victim. Examples include logic bombs, Trojan horses, ransomware, viruses, and worms.
Phishing or spear phishing	A digital form of social engineering that uses authentic-looking, but fake, emails to request information from users or direct them to a fake website that requests information. Spear phishing is a phishing exploit that is targeted to a specific individual or group.
Passive wiretapping	The monitoring or recording of data, such as passwords transmitted in clear text, while they are being transmitted over a communication link. This is done without altering or affecting the data.
Spamming	Sending unsolicited commercial email advertising for products, services, and websites. Spam can also be used as a delivery mechanism for malware and other cyber-threats.
Spoofing	Creating a fraudulent website to mimic an actual, well-known website run by another party. Email spoofing occurs when the sender address and other parts of an email header are altered to appear as though the email originated from a different source.
Structured Query Language (SQL) injection	An attack that involves the alteration of a database search in a web-based application, which can be used to obtain unauthorized access to sensitive information in a database.
War driving	The method of driving through cities and neighborhoods with a wireless-equipped computer (sometimes with a powerful antenna) searching for unsecured wireless networks.
Zero-day exploit	An exploit that takes advantage of a security vulnerability previously unknown to the general public. In many cases, the exploit code is written by the same person who discovered the vulnerability.

Source: U.S. Government Accountability Office, 2015. Cyber Threats and Data Breaches Illustrate Need for Stronger Controls across Federal Agencies. U.S. Government Accountability Office, Washington, DC, p. 5.

Table 5.3 Recent Large Commercial Cyberattacks

Date of Attack	Organization	Records Exposed	Attack Summary
10/3/2013	Adobe Systems, Inc.	152 million	Hack exposed customer names, IDs, passwords, and debit/credit card numbers
12/18/2013	Target Corporation	110 million	Hack exposed customer names, email addresses, home addresses, phone numbers, and credit/debit card numbers
5/21/2014	eBay, Inc.	145 million	Hack exposed customer names, passwords, email addresses, home addresses, phone numbers, and dates of birth
9/2/2014	The Home Depot	109 million	Hack exposed the details from 56 million payment cards and an additional 53 million customer email addresses

Source: Risk Based Security, 2016. Data Breach Quick Review: 2015 Data Breach Trends. Risk Based Security, Richmond, VA.

Risk Based Security, a cyber research group, publishes an annual report of global cybersecurity.[9] Findings for 2015 included:

- 3930 incidents reported which exposed 736 million records
- Of these incidents 2540 were due to hacking. A relatively small amount was due to mishandling, viruses, phishing, etc.
- 40.5% of the incidents and 64.7% of the records were United States-based
- The industry breakdown by incident is: 47.2% (business), 12.2% (government), 13.9% (education), 6.8% (medical), and 19.9% (unknown)
- The industry breakdown by number of records is: 51.4% (business), 12.8% (government), 1.5% (medical), and 34.3% (unknown)

The United States is, by far, the country with the most cyberattacks. Risk Based Security also notes some of the largest commercial cyberattacks in recent times as shown in Table 5.3.

---

[9] Risk Based Security (2016).

The cyberattack on Target is particularly instructive. Hackers got their toehold into the company by attacking one of its vendors, Fazio Mechanical, a small company which provided some of Target's heating and air conditioning. By installing malware in Fazio's computers via an email, they were able to exploit the vendor's link to Target's computer network to push the malware further into Target's computers and computerized cash registers. Soon after the attack was made public many retail firms formed an information sharing organization called the Retail Cyber Intelligence Sharing Center (R-CISC). Memberships include the Retail Industry Leaders Association, American Eagle Outfitters, The Gap, J. C. Penney, Lowe's, Nike, Safeway, Target, VF Corporation, and Walgreen's. The idea is to create a shared situational awareness. This will not fully correct the market failure noted above but it is a step in the right direction.

Cyberattacks on the US power grid are not uncommon, and they follow patterns similar to attacks on commercial supply chains; that is, attackers are looking for entry points along a vast interconnected network. Some recent attacks reported by the U.S. GAO include:[10]

- "Stuxnet" was an attack in 2010. It targeted control systems used to operate industrial processes in various energy and nuclear plants. It was designed to exploit vulnerabilities to gain access to its target and modify code.
- In 2003, the Microsoft SQL Server worm known as "Slammer" infected a private computer network at the idled Davis-Besse nuclear power plant in Oak Harbor, Ohio. It disabled a safety monitoring system for nearly 5 h. The plant's process computer failed, and it took about 6 h for it to become operable again.

Cybercrime, relative to physical theft, has at least five attractions for criminals:

- It is relatively cheap. It is just a matter of learning how to become a computer hacker. Ironically, the Internet has many websites offering how-to advice.
- It is relatively easy. Executing the crime can take place from the comfort of the hacker's hideout.

---

[10] U.S. Government Accountability Office (2012, pp. 10–11).

- It is fast. A lot of sensitive information is stored on files whose size is not prohibitive when it comes time to transfer or copy them.
- It is nearly invisible. Sophisticated hackers can make sure they have removed any digital identifiers thus making it hard to trace the attack back to its source.
- It is easy to escape. In fact, if the hacker is in another country (especially one with spotty cybersecurity laws), there is little chance of prosecution in the jurisdiction where the crime was committed.

Cybercrime has some unique aspects to it. One is that the criminals may be representatives of enemy governments. They may be called cyber-warriors. In this way it is more espionage and warfare than theft for its own sake. Another is that the criminals are doing it in order it to make a political point. In this way they are "hacktivists." Finally, they may be doing it because they derive pleasure in testing the security of the system they are targeting as against their computing prowess. In any case, supply chain partners expect to receive communications from one another. A hacker, whatever the motive is, may wish to target these channels of communication in order to impersonate a partner.

Protecting Information

Computer hardware which stores sensitive information needs to be secured. This means that access to computer stations must be granted with keys, identity badges, biometrics, etc. Once access is granted, it should be further limited based on passwords—cleverly designed ones which are frequently changed. The hardware needs to be in a secure locale ideally away from the perimeter of the organization's property. Only approved software can be uploaded to the computer network. Computer systems need to be backed up and periodically scanned for viruses and other malware. Software can be used to monitor incoming and outgoing emails and Internet searches. Data can also be encrypted; that is, an algorithm is used to scramble the data rendering it useless to anyone who does not know the protocol to unscramble it. Just as with physical theft, criminals look for targets of opportunity. The more hurdles a hacker has to jump over, the more likely it is that the hacker will search for easier marks.

Just like the concept of supply chain management, with its emphasis on trust and breaking down silos, the same philosophy applies to protecting information. Executives know where the information is and what its value is. The IT personnel can provide technical solutions; but they need to be instructed where and when to apply them. Communication in-person and communication online are simply two ways to communicate and both need to be managed. Managers would never let carpenters and decorators design office spaces and meeting areas on their own without their input and approval. It is a matter of office functionality and organizational structure—and the office space is a means to an end. Yet managers often leave it to their IT personnel to decide on much of the computer hardware and security software used. But without their interest and input, the virtual office will not perform well in cyberspace. Executives at the operational level and at the information level (i.e., COOs and CIOs) need to work out common strategies. Basically, it is a matter of discussing business decisions in the context of vulnerability and security of both information and physical space. If a cyberattack were directed to a port, would it disable email communications or would it disable the gantry cranes? Both are possible; therefore, cyberattacks cannot be left only to the IT personnel. In fact, the damage from a cyberattack might be mitigated by physical methods. "For example, there's a simple physical way to prevent tanks from a liquid bulk terminal from overflowing via a cyber hack: Install a lever with a float that acts as a flow switch."[11]

Cybersecurity Strategy

Effective supply chain management also requires that partners speak the same language and have the same intentions. For example, vendors and customers should share the same meaning of product quality; buyers and sellers should share the same meaning for what constitutes on-time delivery; inventory managers should coordinate warehouse utilization with production managers' needs for the assembly line. What should the COO and CIO agree on? Four critical questions to reach agreement on are:

- Where are the organization's most critical assets (both tangible and intangible) to be stored?

---

[11] Szakonyi (2015, p. 12).

- What are the most likely cyber-threats and how would an attack take place?
- What are the IT vulnerabilities?
- What are the consequences of an attack and what is the plan to recover from it?

The next step after securing agreement is for the COO and CIO to set up cross-functional teams to assess risks and measure costs. Their advice back to the COO and CIO will help them decide on an appropriate budget for information security. This process needs to be continuous since IT is subject to rapid change and the organization's trade-offs may be reassessed.

Despite the risks, global supply chain partners have not developed clear industry-wide standards for how to deal with a cyberattack. The federal government has responded with some guidance and invited industry input to get ahead of the hackers. One such program is managed by the U.S. Department of Commerce's National Institute of Standards and Technology. Five core functions are suggested in Table 5.4.

Given the categories shown, the organization and its supply chain partners must rank them in terms of priority; and then identify the gap between how a category is currently working relative to where it needs to be. This will be a theme in the discussion of "as is" processes becoming "to be" processes in enterprise resource planning (ERP) discussed ahead. Feedback from the operational level should inform strategic and budgeting decisions.

LINKING, TRACKING, AND SENSING TECHNOLOGY

Linking Organizations Through Information Technology

Supply chain managers have been working for decades to seamlessly link their upstream suppliers and downstream customers. Technology has been a driving factor. While the goal is to create an information trail that tracks all inputs, all end products made from those inputs, and all deliveries of those end products to their final customers, most supply chains are not that information-intensive nor well-connected.

While supply chain management has always been about breaking down silos, it is ironic that many of the most prominent connectivity programs were in their own silos. Programs such as Material Requirements Planning, Distribution Requirements Planning, and Just-in-Time rose to prominence in the 1980s. Their focus was on information related to

Table 5.4 Cybersecurity Framework Functions

Function	Category
Identify	Asset management
	Business environment
	Governance
	Risk assessment
	Risk management strategy
Protect	Access control
	Awareness and training
	Data security
	Information protection processes and procedures
	Maintenance
	Protective technology
Detect	Anomalies and Events
	Security continuous monitoring
	Detection processes
Respond	Response planning
	Communications
	Analysis
	Mitigation
	Improvements
Recover	Recovery planning
	Improvements
	Communications

Source: National Institute of Standards and Technology, 2016. Framework for Improving Critical Infrastructure Cybersecurity. U.S. Department of Commerce, Washington, DC. http://www.nist.gov/cyberframework/upload/Cybersecurity-Framework-for-FCSM-Jan-2016.pdf.

materials for production and the distribution of end products. These are logistics programs designed to facilitate production. In the early 2000s, Customer Relationship Management became prominent. The focus was on information related to customer behavior. This is a logistics program designed for marketing and customer fulfillment. Another program, ERP, was developed in the 1990s with the goal of getting a company's own departments (i.e., purchasing, marketing, accounting, human resources, etc.) to improve its intra-company coordination of information. It is not uncommon for a large company to store the same data in multiple databases across the company and enter the same data multiple times and by different people into these different database silos. Such databases were often designed to be separate because of the different functional requirements of the departments, or the cultural or technological differences of company subsidiaries in other countries. The result was that it was impossible to perform any intricate or broad-based analyses of how products and information were moving across the company. ERP represents a single database approach. Data would be entered just once and all departments would have complete visibility.

From a supply chain security standpoint the idea of moving the point of compliance away from the US port of entry means CBP and the shippers and carriers under their jurisdiction have to consider the upstream supply chain of the imported item. ERP offers the option to do this effectively; and it could be tied into ACE. However, just like ACE is having problems establishing true connectivity across all 48 partner government agencies (PGAs), ERP implementation in the private sector is not always seamless. Perhaps the largest impediment is the change from managing the information flow from silo to silo, with its own particular valuations and use of the information, to one of a company-wide process. This means mapping the flow process and each department manager understanding his role within that process. Since this can be a culture shock to those employees used to their own legacy systems, upper management needs to spend the time necessary to secure their buy-in. Before moving to the ERP system it may be necessary to run it with legacy systems in place to act as backup while any bugs in the system are being worked out. This has certainly been the practice with the piecemeal roll-out of ACE.

Once ERP is implemented within a company, others can be brought in through electronic data interchange which, in effect, connects each company's computer networks. A quicker option is through the Internet using Cloud computing (which, of course, makes the system as vulnerable as the preexisting software chosen from the Cloud). In either case this is known as ERP II. The intent is to give intercompany visibility and increase the speed of uploading and downloading of information across the various companies.

Simply put, ERP II process mapping involves reengineering the process from an "as is" scenario to a "to be" scenario.[12] The idea is to lay the

---

[12] For a complete outline see Okrent and Vokurka (2004, pp. 641–642).

technology over a more streamlined system since speed is one of the attributes of this technology. Using the technology within a poor process will not improve the process at all. "As is" for, say, how an import is ordered and delivered could be modeled in the following way:

- Gather all the participants together. These would include:
  - The buyer and seller of the import, one of whom will agree to be the consignor who arranges for transportation.[13]
  - The carrier or carriers (if the transport is intermodal) and the freight forwarder if an intermediary is used to represent the consignor.
  - The customs broker if the importer chooses to hire one.
  - A representative from CBP familiar with the screening and inspection processes.
- The participants would produce all paper or electronic documents and computer screenshots involved in handling their activity regarding the import. These are placed in chronological order for all to consider. This would include:
  - Proof of transfer of funds and receipt between the buyer and seller.
  - The bill of lading between the consignor and carrier with attention paid to how, where, and when the transport will take place. There may be multiple bills in an intermodal setting.
  - A document of the services the broker is likely to provide (i.e., tariff classification, payment of duties, etc.).
  - The cargo manifest presented to CBP. Recall from Chapter 4 that Importer Security Filing data may be required as well.
- Connect the documents together with notes concerning the participants responsible for their completion and cycle times necessary to complete the actions mandated by the documents.

Naturally, this process can be even more complicated if the importer imports various types of goods from multiple countries, and uses different carriers and brokers for these. Of course, this type of process mapping can take a considerable amount of time to complete.

---

[13] One could go even deeper and include the buyer and seller's individual banks as well. This applies when the transaction is completed through a letter of credit or a documentary collection sale. The banks would play a role as to when payment is received and the import is released to the carrier. For a detailed outline of these two payment methods see Wild et al. (2006, pp. 378–381).

When the "as is" scenario is mapped, the next step is to map a "to be" scenario. What is involved here is to identify which connections in the "as is" scenario are critical to the process concerned. In the import example, it could be the use of a particular carrier or broker because of their specialized knowledge. It could involve use of a particular route as proscribed in the bill of lading because of the time sensitivity of the shipment. In other words, these connections must fulfill a strategic goal. The next step is to look for "as is" processes that can be eliminated or streamlined because they are not value-adding. After this process has been slimmed down, the ERP system software would be used in an attempt to automate the processes that the various participants are comfortable with. The idea is to use human intervention only when constraints to the system change. In the import example, this could include: a port shutdown, a sharp rise in freight rates or import tariffs, a change in the "24 hour rule" for ocean vessel document submission, etc. In other words, ERP is just as much a human resource and organizational system as it is an IT platform.

Once the "to be" scenario is worked out, the ERP system software can be applied across the companies. However, since ERP software vendors will likely have to customize the product to fit the unique characteristics and needs of the supply chain concerned, there is a risk that security gaps might be created. This opens the supply chain to cybercrime along the lines noted above. This is especially true if transactions are Internet based.[14]

Tracking in Motion

Bar codes dominated the tracking technology industry for decades. They were limited, however, because the code had to be physically scanned over a laser-emitting reader. Radio frequency identification (RFID) tags (which contain antennae) improved the process since the tag simply had to be near a signal reader. These passive tags have a cousin in the form of an active RFID tag which contains an antenna and a battery capable of sending out a signal of its own. RFID tags have been used by pharmaceutical companies such as Pfizer and GlaxoSmithKline since 2006 in order to deter counterfeit items entering their supply chains. Tagging tamperproof bottles is a way to track them as they move downstream

---

[14] Wolden et al. (2015) discuss the cyber-vulnerabilities of ERP and how to mitigate them.

along the supply chain.[15] However, an inherent vulnerability in RFID is that information is sent through the air space between the tag and the reader, and there is no guarantee that the signal cannot be intercepted. Furthermore, since the system owner cannot monitor what is happening in this open space he cannot know when the system has been breached. The signal contains an identification of the item in transit. One way around a security breach is to create a process which randomizes the signal between the tag and the reader and resolves it through a mutual authentication protocol.[16] In this way the cyber-attacker cannot understand the information being transferred.

Global positioning systems (GPS) can certainly shed some light within the black box of transportation. The ability to know where a conveyance is at any moment in time is valuable when it comes to designing transportation networks to facilitate procurement and distribution of cargo. What can be tracked can be controlled to some degree; and this may help mitigate cargo theft as well. If the trailer or container has GPS, it may also have software which provides geo-fencing (i.e., virtual boundaries). The transport route can be defined with a radius of deviation allowed along the way. When the radius is breached, an alarm would be triggered. Taking this one step further, if the trailer/container is being handled by a truck the dispatcher could use technology to remotely shut down the vehicle's engine. RFID readers can be used to set up geo-fencing as well, but they rely on physically placing the readers at the desired boundaries. GPS allows for more flexibility since the virtual boundaries are tracked via a satellite.

Where does this technology leave the criminals? It limits the time they have to separate the cargo from the vehicle and get the former to a safe place. If they have the expertise they can try to dismantle the GPS in order to have more time to maneuver the vehicle; but that means tinkering with the vehicle in the open along the route or within the predetermined radius. Of course, GPS is useless if the operator has an accomplice. In this case it is just a matter of stopping the vehicle en route and handing over the cargo (usually an extra load surreptitiously added at the point of origin).

---

[15] For a review of the companies' decisions and the drive for a government mandate, see Wyld (2008).

[16] Liu et al. (2012) outline a mutual authentication protocol for passive RFID tags.

Sensor-Based Logistics

Tracking items to know where they are in the supply chain is not a new phenomenon. Telegraphs, radios, or telephones could be used to note the approach of a conveyance by land, sea, or air. GPS can now track conveyances in real-time. Today shippers would like information on individual shipments—and so would the carriers and government regulators (for quite different forms of planning). In addition to location, there is value in knowing the temperature of the shipment while in transit. This is particularly important for perishable items such as food, pharmaceuticals, medical specimens, and live animals. Exposure to light is an important piece of information as well, because it can indicate tampering and/or theft. Shocks, vibrations, humidity, and barometric pressure could be included if the entire environmental picture is desired. The technology exists to provide all of this sensory data along with software to display and store the data collected.

Sensor-based logistics (SBL) is what these parties are looking for. They would like to have full visibility in real-time, have the option for geo-fencing, and have the collected data transmitted to all authorized partners to insure collaboration and respond if a problem arises. Of course, the technology has not progressed that far as yet. The limitations come in when one insists on constant updating as opposed to time intervals. The devices used in SBL have batteries with a few days or perhaps a week of power. Also, even if the sensor is updating what it is sensing it may not be able to transmit the data until the conveyance is out of a dead zone with no cell phone, Internet, or GPS coverage. Nonetheless, when an alert is received by an SBL device, intervention time is reduced. This could involve re-freezing "cold chain" cargo, repackaging damaged items, or reporting a suspected theft.[17]

---

[17] FedEx, for example, offers a proprietary SBL known as SenseAware. Launched in 2009, it works like a 3G cell phone and has all the benefits, as well as limitations, noted in SBL. However, it appears attractive to many shippers because the device may be used with other carriers (since the shipper pays FedEx a fee for usage) and the tracking is still handled by the FedEx global tracking system. Given that it uses a lithium battery, the FAA had to approve its presence in the cargo areas on commercial cargo and passenger flights. Finally, FedEx had to work with various customs agencies around the world to make sure that the device is authorized and can enter and exit the country concerned. Currently, eligible countries are: Canada as well as the European countries, Belgium, Czech Republic, Denmark, Finland, France, Germany, Ireland, Italy, Netherlands, Norway, Poland, Spain, Sweden and Switzerland.

THE BIG DATA CHALLENGE

Big Data

The amount of data available to businesses and consumers today is mind numbing. McKinsey Global Institute estimated that in 2010 businesses stored 7 exabytes of data on their computer networks while consumers stored 6 exabytes on their computers and smart devices. Exabytes are one quintillion (i.e., 1×10^{18}) bytes. In terms of size just 1 exabyte is 4000 times the amount of information held by the U.S. Library of Congress.[18] Furthermore, this dataset is expected to grow exponentially in the years to come.

The term "Big Data" is somewhat of a misnomer. Ever since the Renaissance we have lived in a world which has sought answers to scientific questions and which has worked to improve one's life. Science is anything which lends itself to measurement. So, there has been a lot of measuring and gathering of data going on. Even before the modern age of computers there was a lot of data in existence; but it was very costly to try to gather together a lot of it. Now, with the Internet, practically anything uploaded anywhere can be gathered (i.e., searched for online and then downloaded) to anywhere in the world just so long as it involves a computer network. Big Data, therefore, implies that a large volume of data is generated and made easily available to a large number of people through the velocity of computer networks. This makes it tempting to leverage the data to try to solve more problems, with greater precision, and at faster speeds. Of course, as noted earlier, data is not always information. Information is what is used in order to make decisions. So, when it comes to data one must be discriminating. Why? Because one gets closer to "truth" or at least improves one's decision making the more information one has. Since time for making decisions is limited, one must have a sense for how to filter out the noise from the data which truly signals its usefulness.[19]

Data can come in three forms: structured, unstructured, and semi-structured. Structured data is organized because it was gathered through

---

[18] McKinsey Global Institute (2011, p. 3).
[19] This idea of raising the signal-to-noise ratio from a given data set is expressed well in Silver (2012). For a cautionary tale from Ford Motor Company concerning how those who insist on evermore data may be simply encouraging biased data being generated by their employees, see Mayer-Schonberger and Cukier (2013, pp. 164–166).

a database designed to make the data searchable in ways the user finds valuable. The data is held in fixed fields like in a computer spreadsheet. This includes, for example, the number of items taken out of inventory, the corresponding time, and where the items are being sent. It also includes data from on-board recording devices such as computer readings on engine performance and fuel economy. Unstructured data is that which is not held in specific fields. These include emails, digital images and video, call center recordings, and social media posts. There may be valuable information to be gleaned, but the target information could have been expressed or shown in a variety of ways; hence, unstructured. This means that pattern recognition must be made more flexible when dealing with unstructured data. Semi-structured data is not stored in fields, but it does have tags or markers to identify sections of the data. An example would be an electronic book in HTML format. In summary, Big Data has the attributes of volume and velocity, as well as variety.

An avalanche of data is, indeed, available on the Internet. Companies which use barcodes, RFID tags, and GPS systems can generate their own private avalanche of data. Supply chain partners, with their own big datasets and ERP II processes, will only compound the problem for the company. Each organization can process data differently and this can impede insight. For example, one purchase order by a retailer does not translate into one activity by a manufacturer. There may be multiple inventory and production requests to produce the finished product. This may involve multiple vendors, multiple carriers, and multiple ports of entry. As to the ports of entry, collecting data on customs clearance times would be helpful. If the supply chain passes through areas prone to some of the natural threats noted in Chapter 3, it may be necessary to apply weights to the data in order to account for acts of God or assign probabilities based on data from weather services, insurance companies, etc. The idea is to consider Big Data like an ocean being fed into by rivers of data from multiple streams. How deep is the firm willing to dive in order to understand the undercurrents? Data can be mined in order to look for patterns, which proceed upstream to the problem concerned.

What Should Be Analyzed?
Regarding international trade CBP's proprietary targeting algorithm, the Automated Targeting System, takes in data from bills of lading and

cargo manifests and screens it in order to determine which conveyances and shipments warrant further attention. These items include:

- Shipper, carrier, and intermediary histories
- Nature of the items in the shipment
- Country of origin and transshipment (including particular ports)
- Deviations from the norm

The last point means that if the trade pattern and business process do not fit with past practices, or with current intelligence gathered, they raise red flags. While shippers and carriers are required to use reasonable care in their business, all it may take is a mislabeled package or incorrect entry of a document for CBP to raise concerns. Recall, this risk-based strategy is not sufficient to meet Congress' requirement in the SAFE Port Act (2006) of 100% container scanning; but DHS has been able to secure deferrals of this mandate in the years since.

Given today's Big Data world, one could conceivably analyze anything there is enough data on. Computers today certainly are up to the job. But is that efficient? Not from an information gathering perspective. A computer can bark up every stem of a decision tree in order to search for a good decision, but these are only tactics. Human beings can still see the bigger, strategic picture. This is especially important in games which are subject to changes in states of nature. After all, most data collected is already from the past, and the decision maker must deal with the present or the future. However, data from the present can and should be used by business and government to help in decision making. For example, satellite imaging (on clear weather days) can track infrastructure usage, gauge congestion, and inform alternative routing. This combined with analysis by police of social media chatter, credit card and cell phone usage, etc., may improve forecasting.

If data analytics is used, one must take care in making policy based on the predictive factors the algorithm says are most important. This can be controversial in, for example, criminal justice and in insurance. The setting of bail amounts and insurance policy rates, respectively, could be based on the locale or demographic of the "bad" people. Some would call this profiling. CBP, however, has more political cover because entry of visitors and imports is not a right.

An important human quality is the ability to use heuristics. This means a plan which is not optimal but simply good enough given the circumstances. For example, are Customs Trade Partnership Against Terrorism (C-TPAT), Free and Secure Trade (FAST), and the Container Security Initiative (CSI) working well as supply chain security programs? Certainly a lot of data is generated and provided to CBP. But a good heuristic would be: What is the rate of change in those volunteering to join? How many shippers and carriers are C-TPAT members? How many carriers are choosing to use FAST lanes? How many foreign countries are signing up their ports to be in CSI? If the membership is rising, that indicates value; and if it is stagnant or falling, there is likely something wrong. How can we be sure? Consider C-TPAT. Since shippers and carriers interact with each other and each belongs to special interest groups and industry associations, it is likely that a lot of information is shared among C-TPAT members and non-members. From this statistic one could then drill deeper into particular ones if one wished to find out why, for example, membership were falling in a particular region.

Another important human quality is the need to see the big picture. In other words, it is not about doing some task for its own sake; it is because it serves some purpose. The whole is more valuable than the individual parts. This Gestalt theory helps explain why trust and information sharing takes place along supply chains made up of partners who need not cooperate. However, they choose to. In game theory Schelling's focal point is an excellent example of trying to bring order out of chaos and trying to see the big picture when presented with seemingly unconnected options.

ISSUES AND PROBLEMS GOING FORWARD

Since terrorists wish to attack an economy, it is no stretch of the imagination for them to want to target the Internet. E-commerce is under the threat of cyber-terrorism. Any online system is vulnerable to hacking. Disruptions can run the range from identify theft all the way to shutting down physical infrastructure. An attack might take the form of implanting computer viruses or malware in order to spy on Internet communication and, when the moment is right, shut down the website. The websites may control e-commerce or even physical infrastructure such as power grids, water and sewage facilities, etc.

What can be said about an ERP II process mapping for customs compliance? One is that it is quite different from a private sector mapping.

CBP and 47 other PGAs may collaborate with the shippers and carriers; but they may also be restrictive and this differentiates them from other supply chain partners. There is no doubt that government agencies do not have the same incentive to create process maps like private sector companies do. Regulations are formed by the bureaucracy based on its legislative intent. These interpretations may change over time as well. Of course, the process is not one-sided. The private sector does have a voice. Shippers and carriers participate to some extent. Chapter 6 explores how they might influence the way regulations are enacted and thereby affect the process map.

Bibliography

Ackoff, R.L., 1989. From data to wisdom. Journal of Applied Systems Analysis 16, 3–9.

Barney, J., 1991. Firm resources and sustained competitive advantage. Journal of Management 17 (1), 99–120.

Brandon-Jones, E., Squire, B., Autry, C.W., Petersen, K.J., 2014. A contingent resource-based perspective of supply chain resilience and robustness. Journal of Supply Chain Management 50 (3), 55–73.

Kelly, E., October 20, 2014. Officials Warn 500 Million Financial Records Hacked. USA Today.

Liu, An-T., Chang, H.K.-C., Lo, Y.S., Wang, S.-Y., 2012. The increase of RFID privacy and security with mutual authentication mechanism in supply chain management. International Journal of Electronic Business Management 10 (1), 1–7.

Mayer-Schonberger, V., Cukier, K., 2013. Big Data: A Revolution That Will Transform How We Live, Work, and Think. Houghton Mifflin Harcourt Publishing Company, New York, NY.

McKinsey Global Institute, 2011. Big Data: The Next Frontier for Innovation, Competition, and Productivity. McKinsey & Company, New York, NY.

Mellat-Parast, M., Spillan, J.E., 2014. Logistics and supply chain process integration as a source of competitive advantage. International Journal of Logistics Management 25 (2), 289–314.

National Institute of Standards and Technology, 2016. Framework for Improving Critical Infrastructure Cybersecurity. U.S. Department of Commerce, Washington, DC. http://www.nist.gov/cyberframework/upload/Cybersecurity-Framework-for-FCSM-Jan-2016.pdf.

Okrent, M.D., Vokurka, R.J., 2004. Process mapping in successful ERP implementations. Industrial Management & Data Systems 104 (8), 637–643.

Prentice, B.E., Prokop, D., 2016. Concepts of Transportation Economics. World Scientific Publishing Co, Singapore.

Risk Based Security, 2016. Data Breach Quick Review: 2015 Data Breach Trends. Risk Based Security, Richmond, VA.

Silver, N., 2012. The Signal and the Noise: Why So Many Predictions Fail—but Some Don't. The Penguin Press, New York, NY.

Steptoe and Johnson LLP, 2016. Comparison of US State and Federal Security Breach Notification Laws. Steptoe and Johnson LLP, Washington, DC. http://www.steptoe.com/assets/htmldocuments/SteptoeDataBreachNotificationChart.pdf.

Szakonyi, M., April 20, 2015. Plugging port holes. Journal of Commerce 10–14.

Urciuoli, L., Hintsa, J., Ahokas, J., 2013. Drivers and barriers affecting usage of e-customs: a global survey with customs administrations using multivariate analysis techniques. Government Information Quarterly 30, 473–485.

U.S. Government Accountability Office, 2015. Cyber Threats and Data Breaches Illustrate Need for Stronger Controls Across Federal Agencies. U.S. Government Accountability Office, Washington, DC.

U.S. Government Accountability Office, 2012. Cybersecurity: Challenges in Securing the Modernized Electricity Grid. U.S. Government Accountability Office, Washington, DC.

Wild, J.J., Wild, K.L., Han, J.C.Y., 2006. International Business: The Challenges of Globalization. Pearson Education, Inc., Upper Saddle River, NJ.

Wolden, M., Valverde, R., Talla, M., 2015. The effectiveness of COBIT 5 information security framework for reducing cyber attacks on supply chain management system. IFAC-PapersOnLine 48 (3), 1846–1852.

Wyld, D.C., 2008. Genuine medicine? Why safeguarding the pharmaceutical supply chain from counterfeit drugs with RFID is vital for protecting public health and the health of the pharmaceutical industry. Competitiveness Review 18 (3), 206–216.

CHAPTER 6

The Business and Government Interface

CONTENTS

THE GOALS OF BUSINESS

Private Goals

Unlike government, with its power to print money, raise taxes, and issue bonds, businesses face a financial bottom line. They face pressure to control costs (unless they are a monopoly) and to provide a product wanted by their customer base at a price which at least covers all operating costs incurred in making the product. This is the heart of what is known as productive efficiency and allocative efficiency, respectively. Since businesses can go bankrupt, it is necessary for them to have a strategy to survive and, hopefully, thrive in the marketplace. Cost control, relative to revenue generated, is one way to try to increase profits.

119

Global Supply Chain Security and Management. http://dx.doi.org/10.1016/B978-0-12-800748-8.00006-6

Another possibility is to increase market share in the hope of growing revenues faster than costs. Finally, publicly traded firms may feel the pressure to increase shareholder value which, of course, is dependent on the attitudes of the stock market. The market may reward the firm for revenue growth or it may prefer cost control.

No matter which private goal the firm opts for, it will have to formulate strategy and conduct its business within a tangle of laws and regulations. Some of these may help the firm and some may hinder it. Of course, this does not mean that the firm is insensitive to the goals of government or of the society in which it exists. Many firms have long recognized the importance of being what is often called good corporate citizens. However, it is a balancing act to serve a narrow market of producers and consumers while being aware of the shifting politics of society at large.

Social Goals

Beyond simply serving a market for their own products, many firms feel the pressure or the responsibility to serve the public at large. Concepts such as environmental sustainability, corporate social responsibility (CSR), and ethical business practices—which are all related to one another—have made it into strategic discussions at the boardroom level. Private goals are much more objective and easier to measure since they tend to be dollar-based. Social goals are more nebulous because they are harder to measure and much more subjective. The definition of the word sustainability itself does not enjoy a consensus.[1] A simple definition is doing business to meet the needs of the present without diminishing future generations from fulfilling their own needs—a lofty goal. Nonetheless, many firms have used social goals to try to enhance their private ones. For example, Starbucks' purchase of "fair trade" coffee beans sends a signal to its customer base, which, presumably, wants its coffee along with a feeling that the distant farmers are not harmed economically as a result.

CSR may be defined as: "a company's commitment to minimizing or eliminating any harmful effects and maximizing its long-run beneficial

---

[1] Giunipero et al. (2012, p. 260). Their own definition of supply chain sustainability is: "the extent to which supply management incorporates environmental, social, and economic value into the selection, evaluation and management of its supply base." Of course, these three values—as a sort of triple bottom line—may be at odds with each other.

impact on society"[2] The idea is to improve the company's reputation and thereby increase a consumer's willingness to buy, an employee's willingness to work, and an investor's willingness to invest. The firm is thinking beyond its narrow economic and legal obligations. From an economic perspective CSR puts some or all of the responsibility of eliminating negative externalities on the firm itself. But a criticism is that doing so can take the firm off of its private goals to the point that its cost of doing business becomes too high to sustain growth. If these externalities affect society, so goes the critique of CSR, then let government handle them via taxes and regulations and be answerable to the electorate if the policies work or not. In other words, firms will look after their bottom lines and government will incentivize their "good behavior." From a supply chain perspective, things are more complicated because a firm and its vendors may have different views on CSR and each plan may not work well with others (e.g., a US retailer which imports goods from a vendor in a country which uses child labor). The idea is for the firm to promote a congruous form of CSR where items move downstream along the supply chain and all parties agree on what value (economic and social) is to be added along the way.[3] They should, therefore, develop a common definition of sustainability. Of course, this is even more challenging when supply chains are international and environmental laws can be so different.

Supply chain security fits quite well into both private and social goals of businesses. In a private sense a disruption of the supply chain will cost the firm both time and money. It would ill-affect any private goal the firm might have. Therefore, it is important for the firm to invest in a supply chain security plan. In a social sense a disruption of a firm's supply chain, if it is serious enough, has the potential to raise insurance rates across the industry because of the precedent set. If the incident occurred along public infrastructure it may be closed down by the authorities thus causing delays and congestion for many more firms. If it is a terrorist incident the public's sense of panic is not likely to focus just on the targeted firm. In other words, firms want to see secure supply chains as much as the general public does. However, the challenge remains: If CSR is part of a given supply chain will it be congruent along it? C-TPAT, when a

---

[2] Mohr et al. (2001, p. 47).
[3] For an examination of congruous CSR, see Hietbrink et al. (2010).

shipper and its carrier are both members [meaning Customs and Border Protection (CBP) approves of their plans], offers a chance for congruous supply chain security. Of course, the shipper's upstream supplier should ideally be in C-TPAT or have similar plans in place. While C-TPAT is a government-initiated program, the private sector does have a role to play in how it evolves through a particular PPP to be discussed below.

Some goals cannot be met by private businesses alone; hence, the need for government. Government can gather intelligence, play the role of police officer, and gather taxes to finance these activities. In other words, government can play the role of enforcer or coercer in ways that businesses cannot—and should not—play.

THE GOALS OF GOVERNMENT

Business Partner or Police Officer

Partnering or policing is a choice between carrot and stick and the choice is made exclusively by the government. In fact, there are three choices the government faces:

- Top-down policing: security policies and programs are set by the government and businesses are compelled to follow them. All government interaction with businesses will be in the form of policing–a one-sided relationship.
- Bottom-up outsourcing: security policies are vague and programs, if any, are flexible to the abilities of businesses to implement their own security programs. Policing and enforcement of the law would be reactive to requests by businesses—a one-sided relationship.
- Hybrid collaboration: a public-private partnership (PPP) with shared goals and responsibilities. A two-sided relationship; but asymmetric in favor of the government.

The PPP relationship has the government as, in effect, the senior partner; it is necessary because it has the legal power to police and coerce. If the government overreaches or underperforms, it is answerable to the people at the ballot box.

The Transaction Cost Approach

When considering an appropriate way for business and government to interface, it is useful to consider the transaction costs involved.

Recall that logistics is a transaction cost in that when buyers and sellers agree to a transaction they need to exchange money or credit. Today this can be done electronically and very quickly. But the physical item bought needs to move from point A to B in order to finish the transaction. This cost of transportation is independent of the value that the buyer and seller place on the item. Transportation and logistics, in general, are a transaction cost which is a necessary part of the desire to trade. Supply chain management involves transaction costs as well. An automobile producer, for example, who wishes to buy steel from an appropriate vendor needs to take the time to find and vet the vendor from the pool of competitors. Once done, there is the time involved in negotiating terms of service and enshrining them in a legal contract.

There are two types of transaction costs any organization might face when it buys or sells a good or service. One type are *ex ante* costs such as the construction of the infrastructure necessary to provide security. The other are *ex post* costs which involve monitoring, inspecting, and enforcement. Consider the *ex ante* costs. While the private sector can build airports and ocean ports and run them for profit, it would be more controversial to do the same with land border crossing points. Apart from landing fees for passenger and cargo airplanes an airport's revenue can be earned through vending licenses, sale of fuel, and rents from businesses which lease airport land. Ocean ports, likewise, can earn revenue beyond vessel berthing fees. Port land can be leased and fuel can be sold. But land border crossings are not places where passengers and cargo may wish to linger. In order for the private sector to earn enough revenue at a land border crossing it would have to, in effect, operate like a toll service. Government is better suited to own the appropriate infrastructure. Now, consider the *ex post* costs. Obviously, it is more appropriate for government to handle monitoring, inspecting, and enforcement. But, while government is better able to take on these costs, it does not mean that businesses face no transaction costs of their own. They face the compliance costs to apply security mandates, negotiate with government when allowed, and the up front costs to develop technologies to ease compliance costs. Of course, technology is an area where businesses can help government. The private sector is better adept at allocating capital among risky projects and can advise the government about what is technologically possible.

PPP is a necessary compromise so long as the government wishes to both facilitate trade and provide supply chain security. Policing is the right activity to provide security. But outsourcing security leaves it to the private sector to decide between trade and security; and it is more likely to choose trade over security because there is more profit to be had in the former. PPP forces a collaboration to decide on the mixture.

PUBLIC-PRIVATE PARTNERSHIPS

Definition

PPPs represent an agreement between private and government entities to work together to meet some mutual goal. This is accomplished through following a similar strategy, sharing information, as well as risk. This sounds a lot like the philosophy of supply chain management. PPPs have two differences: (1) while PPPs are designed like supply chains to be mutually beneficial, they are also formed to achieve some social or public good; and (2) if government is also the regulator in the PPP, the relationship will not be even and not subject to the simple contracting process which characterizes standard supply chains.[4]

When it comes to supply chain security, CBP is the lead organization in the PPP; however, with ACE, CBP is becoming a facilitator to potential PPPs with other trade-related agencies. There is no reason for each agency's trade and security agendas to match. This is a challenge for the private sector firms. Since supply chain security is, or ought to be, a part of each partner's long-term business plan, an appropriate definition of PPP should emphasize long-term cooperation leading to a mutual gain.

Intent

Criminals test the resolve of law enforcement. Law enforcement agencies, in turn, try to think like the criminals in order to stop them. Private sector businesses are stuck in the middle. They rely on law enforcement;

---

[4] PPP literature related to supply chain management is still rather sparse. One paper by Davis and Friske (2013) examines PPPs in order to facilitate US-Canada border clearance. Since the paper focused more on infrastructure development as opposed to security, the policing function was not a major factor. As such, some of the government agencies were happy to allow the private sector to take the lead in PPP strategy and operations.

but must comply with its regulations. Of course, if they simply rely on law enforcement it is not likely to be enough to prevent criminal activity. The private sector, as well, needs to think like the criminals in order to make themselves a less inviting target.

With law enforcement and the private sector sharing the goal of crime prevention, it makes perfect sense to try to partner their efforts. As noted earlier the government may be wearing two hats: partner and police. Trade security is facilitated through data supplied by importers/exporters and carriers. Therefore, CBP is dependent on the accuracy and timeliness of the data supplied.

How can private businesses enhance the eyes and ears of law enforcement? While businesses are certainly used to strategic planning to achieve their business goals, few have a lot of experience applying strategic thought to the social threat of terrorism or organized crime. To do so requires an understanding of geopolitical threats and how they change with emerging technologies. Like supply chain management among businesses, the government and their supply chain partners must be willing to share information and jointly develop training programs.

Some Examples

The Federal Emergency Management Agency's (FEMA's) choice of vendors to partner with represents a PPP in order to facilitate humanitarian logistics. The goal is to restore law and order, as well as safety and security. The PPP relationship among the partners is not one of regulation or policing. In this way, it is very similar to a standard business-to-business (B2B) relationship when market forces are driving it. However, as a government entity, FEMA may not feel the pressure that private sector firms do in having to choose appropriate partners (based on, say, cost or quality) and setting up strong contractual relationships. Indeed, the transaction cost approach is noted for the view that contracts can never be written to cover all possible contingencies. This means there is always room for opportunism (or working for self-interested private goals). In the public setting transaction costs may be higher than otherwise.[5]

---

[5] For the differences in ethics and practice of public and private sector contracting see Hawkins et al. (2011, p. 569).

Vendor partnership with FEMA is voluntary and the vendors expect to be paid (unless they are willing to donate their services). To better gauge PPP structure, consider the seven "best practices" proposed by the National Council for Public-Private Partnerships[6]:

1. Public sector champion
2. Statutory environment
3. Public sector's organized structure
4. Detailed contract (business plan)
5. Clearly defined revenue stream
6. Stakeholder support
7. Pick your partner carefully (not necessarily the lowest bid)

In terms of these seven points the ones which are the most challenging for FEMA are (3), (4), and (7) because they go to the heart of whether or not the contract/plan is based on a proactive, reactive, or hybrid approach (as discussed in Chapter 4). However, once this is decided upon everything else becomes more workable.

C-TPAT, as discussed in Chapter 4, is also a PPP; but it is not market-based. Regulations and policing mix in with CBP's interest in the partnership. As a voluntary program, C-TPAT relies on security exchanges between the private and public sectors in order to be successful.[7] *Ex ante* costs may be reduced for government and business members since the security plans are collaborative. The negotiating and contracting process may be more cordial. *Ex post* costs are likely to be reduced since government can inspect facilities beyond US ports of entry and devote more attention to shippers and carriers which are not C-TPAT members. This gives the government a better feel for how secure the supply chain is. Also, monitoring, at least for C-TPAT members, would be more random than obligatory. This is because C-TPAT is designed to give the shippers and carriers faster clearance, which, for their part, lower their own *ex post* costs.

Customs brokers, which can be hired to act as intermediaries between importers and CBP, have their own PPP with CBP. This is intended to

---

[6] National Council for Public-Private Partnerships (2016). "7 Keys to Success." website.

[7] Voss and Williams (2013) refer to the establishment, cultivation, and maintenance of this exchange of information as relational security.

smooth the process of customs clearance. Similar to the voluntary nature of C-TPAT the National Customs Brokers and Forwarders Association of America (NCBFAA) partnered with CBP to form the Broker-Known Importer Program (BKIP). Customs brokers who are members of BKIP share trade intelligence with CBP. More importantly, for new importers or those changing their trade patterns, broker members may vouch for an importer before the item(s) reaches the port of entry. The understanding is that the broker has had an in-depth conversation with the importer-client about all applicable trade regulations. ACE is designed to receive "known importer" transmissions from broker members and, as such, may prevent any red flags from being raised during the import screening process. It is up to the broker to decide which clients, if any, to represent via BKIP. Also, it is the NCBFAA which provides the guidance for what to consider as opposed to CBP. Suggested characteristics of the client include[8]:

- Closeness and longevity of the relationship
- Volume and value of items handled

As such, this PPP is less restrictive than is C-TPAT. In fact, while CBP handles the periodic reviews and inspections of shippers and carriers, it is the BKIP member broker which takes on the role of reviewing the relationship with the importer. CBP appears willing to let the brokers do the vetting since they are considered a trusted source. Recall that brokers are licensed by the federal government and well versed in trade compliance. CBP, however, does request that details relating to the conversations between brokers and known importers be documented and made available on request.

Operations Collaboration: The Advisory Committee on Commercial Operations

Since 1987, the Department of the Treasury (which was the original manager of the customs function) has worked with a PPP known as the Advisory Committee on Commercial Operations (abbreviated COAC). When DHS was created in 2002, and the customs function was transferred to it under the renamed Customs and Border Protection (CBP), COAC has worked with that department. Mandated by the US Congress,

---

[8] The complete list may be found at National Customs Brokers and Forwarders Association of America (2016).

COAC advises on a variety of customs and treasury issues. More specifically, COAC is concerned with:

- Agricultural inspection
- CBP modernization
- Customs broker regulations
- Global supply chain security
- Intellectual property rights protection
- Revenue modernization
- Trade enforcement

COAC meets about four times per year and provides its advice and recommendations to the commissioner of CBP and the deputy assistant secretary for tax, trade, and tariff policy at the Department of the Treasury.

COAC consists of 20 members along with two government-based co-chairs from DHS and Treasury. Technically, the co-chairs are not members of COAC but do preside over all meetings. They manage and participate in all deliberations but do not vote on any COAC actions to provide advice and recommendations. The 20 members, appointed for two-year terms, are from the private sector representing firms which are affected directly by CBP activity.[9] They are chosen such that there is no regional concentration of members and no more than 10 members may be of the same political party affiliation. While the members are from specific firms their membership is to be considered personal; therefore, no alternate or proxy may take the member's place regarding COAC business. In any case, the intent is to have members with hands-on experience who are familiar with the process and much of the paperwork (or electronic submissions) that are involved in dealing with CBP.

COAC may divide its work among subcommittees of its own choosing but all voting must take place over the entire membership. Currently, COAC maintains six subcommittees with the following names and responsibilities:

- One US Government at the Border: generates advice and develops recommendations related to ACE's "single window" as

---

[9] For example, some of the members of the 14th COAC term are from the following firms and special interest groups: Abbott Laboratories, The Airforwarders [sic] Association, Airlines for America, Chrysler Group, Costco Wholesale, DHL Global Forwarding, Expeditors International, Kraft Foods, Microsoft Corporation, OOCL, and the U.S. Chamber of Commerce.

it relates to the development, communications, metrics, inclusion of ACE's PGAs, and ensuring international interoperability. The committee states that its priorities are to "lower costs, increase competitiveness and the ease of doing business, and utilize as much of the existing current system design ensuring that a modern business model is being used for importing goods into the United States."[10]

- Global supply chain: generates advice and develops recommendations pertaining to the safe and expedited movement of cargo through the global supply chain. This relates to North American competitiveness and the removal of trade barriers.
- Export: generates advice and develops recommendations related to export procedures, enforcement, and facilitation issues. This subcommittee also works with committees outside of COAC. These include: the President's Export Council Subcommittee on Export Administration (PECSEA) administered by the Bureau of Industry and Security (BIS), the Defense Trade Advisory Group administered by the Department of State Directorate of Defense Trade Controls, and the Advisory Committee on Supply Chain Competitiveness administered by the International Trade Administration.
- Trusted trader: generates advice and develops recommendations related to CBP's trusted trader program which began a pilot test in 2015. The intention is to meld the benefits of C-TPAT (i.e., supply chain security) with incentives for importers to enhance their trade compliance programs. As yet CBP has not announced specifically what the extra benefits to these low-risk importers might be.
- Trade modernization: generates advice and develops recommendations pertaining to the modernization of CBP's operational and automated support systems. This includes international liaison with other customs organizations.
- Trade enforcement and revenue collection: generates advice and develops recommendations related to improving enforcement of US trade laws, as well as the collection of trade duties and fees.

COAC, like DHS itself, feels the tension between supply chain security and trade facilitation. This can be seen in the divisions of the current

---

[10] See COAC (2015, p. 2).

subcommittees. When COAC first met in 1987, issues of security and enforcement tended to revolve around illicit drug trafficking and goods smuggling. In the post-9/11 world antiterrorism measures have come to the fore. COAC's challenge has been to assist with the roll-out of ACE so that all PGAs are able to use the system, as well as CBP. In other words, trade is not really facilitated via ACE if CBP preclears the shipment while one PGA was not able to process the data as fast and the shipment is then held up on arrival.[11] Another sign of COAC's shift to trade facilitation was its recommendation in 2010 to delay the 100% container scanning mandate within the SAFE Port Act in preference for the current risk-assessment approach.

Research Collaboration: DHS Centers of Excellence

When DHS was created in 2002, it was also mandated to establish centers of excellence in order to collaborate with universities in applicable research. DHS chooses its university partners based on a competitive process. The university upon receiving the award establishes relationships with the private sector, think tanks, and laboratories in order to fulfill its mandate. Currently, the 13 centers of excellence and their missions are:

- Arctic Domain Awareness Center of Excellence (ADAC) led by the University of Alaska Anchorage. The mission is to develop technology, products, and educational programs to improve situational awareness and crisis response capabilities related to the emerging maritime challenges arising in Arctic.
- Center for Advancing Microbial Risk Assessment (CAMRA) co-led by Michigan State University and Drexel University and established jointly with the US Environmental Protection Agency. The mission is to fill gaps in risk assessments for mitigating microbial hazards.
- Center for Maritime, Island and Remote and Extreme Environment Security (MIREES) co-led by the University of Hawaii and the Stevens Institute of Technology. The mission focuses on developing research and education programs addressing maritime domain awareness to safeguard populations

---

[11] This problem was anticipated by a COAC member (from toy manufacturer Hasbro) which had a product imported from China precleared by CBP but held for inspection by the Food and Drug Administration. The complete story is covered in Edmonson (2007).

and properties in geographical areas that present significant security challenges.

- Center for Visualization and Data Analytics (CVADA) co-led by Purdue University and Rutgers University. The mission is to create technologies needed to analyze the large quantities of information generated in order to detect security threats to the nation.
- Center of Excellence for Awareness and Localization of Explosives-Related Threats (ALERT) led by Northeastern University. This mission is to develop the means and methods to protect the United States from explosive-related threats.
- Center of Excellence for Zoonotic and Animal Disease Defense (ZADD) co-led by Texas A&M University and Kansas State University. The mission is to protect the United States' agriculture and public health sectors against high-consequence foreign animal threats and emerging zoonotic disease threats.
- Coastal Hazards Center of Excellence (CHC) co-led by the University of North Carolina at Chapel Hill and Jackson State University. The mission is to perform research and develop education programs to enhance the ability to safeguard populations, properties, and economies from natural disasters.
- Food Protection and Defense Institute (FPDI) led by the University of Minnesota. The mission is to defend the safety and security of the food system by conducting research concerning the vulnerabilities in the food supply chain.
- Maritime Security Center of Excellence (MSC) led by the Stevens Institute of Technology. The mission is to enhance maritime domain awareness and develop strategies to support the marine transportation system's resilience and educational programs for current and aspiring homeland security practitioners.
- National Center for Border Security and Immigration (NCBSI) co-led by the University of Arizona and the University of Texas at El Paso. The mission is to develop novel technologies, tools and advanced methods to balance immigration and commerce with effective border security.
- National Center for Risk and Economic Analysis of Terrorism Events (CREATE) led by the University of Southern California. The mission is to develop tools to evaluate the risks, costs, and consequences of terrorism.

- National Center for the Study of Preparedness and Catastrophic Event Response (PACER) led by Johns Hopkins University. The mission is to optimize medical and public health preparedness, mitigation, and recovery strategies in the event of a high-consequence natural or man-made disaster.
- National Consortium for the Study of Terrorism and Responses to Terrorism (START) led by the University of Maryland. The mission is to provide policy makers and practitioners with findings related to the human elements of the terrorist threat and help make informed decisions on how to disrupt terrorists and terrorist groups.

Limitations and Outsourcing Security

Some shippers and carriers, and even the government, are willing to pay private companies to provide security services. While the intent need not be to replace government-provided police protection all together, it does suggest that some organizations see a net benefit in paying for extra security. Some of the most prominent private security companies (PSCs) include:

- Academi: originally named Blackwater USA. This military-style PSC also contracted with the US government in the 2003 war in Iraq. Among its services are "stability and protection to people and locations experiencing turmoil."
- ADT Corporation: originally known as American District Telegraph, ADT started out as a telegraph company. This PSC specializes in alarm and security monitoring.
- Andrews International: specializes in armed and unarmed security guards. They guard "national landmarks, major tourist attractions, industrial sites, educational and financial institutions, healthcare facilities, and other locations where security stakes and client expectations are high."
- The Brinks Company: synonymous with cash-in-transit and armored car services. This PSC also specializes in international transportation of valuables and payment management.
- Pinkerton Government Services: in addition to security guards, this PSC also provides firefighting and emergency medical services.

Out of all the modes of transportation the ocean vessel mode is unique in that routes can go through stateless areas (i.e., international waters).

It is true that if the vessel is flying a particular country's flag it is considered an extension of that country's sovereign territory. It is also entitled to the protection of that country's navy. Of course, there is no guarantee that such protection is readily available when needed. The discussion of maritime piracy in Chapter 3 showed that such naval guarantees may not deter pirates (especially those operating in international waters near failed states). The International Chamber of Shipping (ICS), which represents about 80% of the world's merchant fleet, changed its policy in 2012 and now supports the presence of private armed guards on vessels in order to repel pirates. Ship owners flying the US flag have the discretion to hire private security companies for this purpose. However, the US Department of State licenses any guards on board and their weapons are restricted to less than automatic. It also recommends that the use of on-board guards be restricted to vessels operating in "high risk waters." The owner of the vessel is required to perform background checks on the guards and the decision concerning when to use force rests with the captain of the vessel. It is also important to note that the guards on board are not considered to be crew members in the eyes of the government; rather, they are considered to be passengers in terms of safety and security planning and customs entry.

The full range of activities the private security companies offer may be divided into four groups[12]:

- Security intelligence, risk assessment, and consulting
- Security services
  - Training crews
  - Vessel escorts
  - Guards on board
  - Training local security forces
  - Acting as local security forces
- Crisis response and hijacking negotiation
- Intervention
  - Liberate hostages of ships
  - Intervention on land

The risks and litigation involved may be considerable. What happens when a pirate, crew member, or guard is injured or killed? What are the

---

[12] A full discussion of these is provided in Struwe (2012).

insurance implications? These are difficult questions to answer given the range of situations that can arise in international waters.[13] One prominent ocean vessel company which operates in waters off of West Africa has a policy that if any of their on-board guards fires his weapon then the security company has failed them and the company will not be re-hired.

ISSUES AND PROBLEMS GOING FORWARD

A willingness to form PPPs to deal with supply chain security ought to increase collaboration (meaning more partnership and less policing on the part of the government). But will it help mitigate risk? How will it affect the apparent trade-off between trade flows and supply chain security? This will have important implications for both business and government. This issue is explored in detail in Chapter 7. In terms of risk mitigation and measuring success it begs the question: What does success look like when it comes to supply chain management? This is the topic of Chapter 8.

Bibliography

COAC, April 17, 2015. Statement of Work for the Subcommittee on One U.S. Government at the Border (1USG). US Customs and Border Protection, Washington, DC. Website https://www.cbp.gov/sites/default/files/documents/One%20U.S.%20Government%20at%20the%20Border%20Subcommittee%20Statement%20of%20Work.pdf.

Davis, D.F., Friske, W., 2013. The role of public-private partnerships in facilitating cross-border logistics: a case study at the U.S./Canadian border. Journal of Business Logistics 34 (4), 347–359.

Edmonson, R.G., Feburary 26 2007. COAC's pendulum swings back: private-sector advisory group asks customs to pay more attention to trade facilitation. Journal of Commerce 38–39.

Elms, H., Phillips, R.A., 2009. Private security companies and institutional legitimacy: corporate and stakeholder responsibility. Business Ethics Quarterly 19 (3), 403–432.

Giunipero, L.C., Hooker, R.E., Denslow, D., 2012. Purchasing and supply management sustainability: drivers and barriers. Journal of Purchasing and Supply Management 18 (4), 258–269.

Hawkins, T.G., Gravier, M.J., Powley, E.H., 2011. Public versus private sector procurement ethics and strategy: what each sector can learn from the other. Journal of Business Ethics 103, 567–586.

---

[13] For a discussion of the ethics and responsibilities involved when a firm hires a private security company see Elms and Phillips (2009).

Hietbrink, J.J.C., Berens, G., van Rekom, J., 2010. Corporate social responsibility in a business purchasing context: the role of CSR type and supplier product share size. Corporate Reputation Review 13 (4), 284–300.

Mohr, L.A., Webb, D.J., Harris, K.E., 2001. Do consumers expect companies to be socially responsible? The impact of corporate social responsibility on buying behavior. Journal of Consumer Affairs 35 (1), 45–72.

National Council for Public-Private Partnerships, 2016. 7 Keys to Success. http://www.ncppp.org/ppp-basics/7-keys/.

National Customs Brokers and Forwarders Association of America, 2016. Broker Known Importer Program (BKIP). Website http://www.ncbfaa.org/Scripts/4Disapi.dll/4D-CGI/cms/review.html?Action=CMS_Document&DocID=16880&MenuKey=about.

Struwe, L.B., 2012. Private security companies (PSCs) as a piracy countermeasure. Studies in Conflict and Terrorism 35, 588–596.

Voss, M.D., Williams, Z., 2013. Public-private partnerships and supply chain security: C-TPAT as an indicator of relational security. Journal of Business Logistics 34 (4), 320–334.

CHAPTER 7

Trade Efficiency and Security: Are They a Trade-off?

CONTENTS

SUPPLY CHAIN REGULATION

The Cycle of Regulation

The history of commercial transportation policy in the United States has completed a circle. From the early twentieth century to today it has gone from no-regulation to regulation, and then from de-regulation to re-regulation. The re-regulation is most pronounced in the areas of the environment and security. The optimal level of regulation is a political choice and it changes over time and is reactive to events. Hence, a full-circle may be a rational process. Economic regulation is also an act of government and is, as such, a political action prone to the changing sentiments of successive governments and the electorate. The economics of regulation may be set in the context of a trade-off. The motor carrier, air, ocean vessel, and rail modes of transportation in the United States have all enjoyed long periods of de-regulation to varying degrees; however, the post-9/11 state of international trade is one characterized by re-regulation and these modes have had to adjust in ways made clear in Chapter 4.

137

Government regulation represents a *service* to businesses. The government lays the groundwork for a business and its supply chain partners to achieve market efficiency and security in a variety of degrees. In a sense, the supply-side of regulation is modeled in terms of the provision of two service attributes: supply chain security and market efficiency. However, it is the behavior of businesses in response to the particular levels of regulation provided that determines the true level of these two attributes. This model is referred to as the security-efficiency (S-E) trade-off.[1] With this model government programs to regulate, supply chains may be seen in a broader context.

Containerization

The shipping container highlights the market efficiency and security trade-off. The modern version of the shipping container grew popular because it avoided costly and laborious repetition when cargo had to be loaded and unloaded when moving from one mode of transport to another, say from a truck to an ocean vessel. Today, it is just a matter of moving the whole container with its cargo inside.[2] Of course, with the cargo locked inside the container at its source, and not retrieved until the container is opened at its destination, it would appear that security is enhanced as well. After all, more steps of loading and unloading increases the risk of cargo damage, theft, and smuggling. However, as the container moves from one point along the supply chain to another, and from one mode of transport to another, the movers are dependent on the honesty and accuracy of the shipping documents accompanying the container. These documents note cargo contents and weight among other things. Before containerization, a carrier knew what was being placed in the truck trailer, the rail hopper car, or the cargo hold simply by observing the cargo loading process. Today, a container vessel captain only really knows how many containers he is carrying; the contents are less

---

[1] The seminal paper outlining the S-E trade-off is Prentice (1994) and further refined in Prokop and Prentice (2009). In these works the model examined the broader area of business stability rather than security. It should also be noted that regulation characterized as a service—distinct from physical production—requires a blurring of the typical distinctions involved between demander and supplier. A principal-agent approach is necessary.

[2] The complete history of the shipping container may be found in Levinson (2006) and Donovan and Bonney (2006).

tangible in his mind. Therefore, a trade-off remains to the extent that speed is precluded by the time necessary to complete and verify all necessary documentation. Inaccurate declarations of cargo contents and cargo weight are still a security and safety issue.

While it is not possible to guarantee complete accuracy of documentation across the millions of cargo containers moving in, within, and out of the United States, mitigation of the problem is possible. Physical inspection is possible at multiple steps, but this would amount to the same thing as unloading and reloading in the era before containerization. For international transport the container must pass through the screening and/or inspection processes of Customs and Border Protection (CBP). Accuracy of documents are necessary in order to determine tariffs and duties. Also, if the container is held in a foreign port within the Container Security Initiative (CSI) program there is the added layer of security upon departure for the United States. For surface transport the C-TPAT program incentivizes shippers and carriers to be accurate.

As to cargo weight, the challenge has been where the weighing of the laden containers should take place: at origin or at port of entry. Inaccurate weights can create safety issues for carriers and crews. Some recent incidents include the *MSC Napoli* in the English Channel in 2007. The hull collapsed when faced with a strong storm. The captain intentionally beached the vessel to prevent further damage. Out of the 660 containers which were transported, 137 (or about 21%) were found to be over their declared weight by a range of 3–20 metric tons. The *Deneb* in the Port of Algeciras in 2011 suffered a similar fate. The ship nearly capsized on its starboard side because 1 out of every 10 containers had declared weights significantly below their actual weight. Sixteen out of 168 containers being, on average, four times overweight was enough to cause the accident. The UK Marine Accident Investigation Branch, following up on a collapse of stacks of containers on the *P&O Nedlloyd Genoa* in 2006, reported: "[I]ncorrect weight can result in stack overload and the application of excessive compression and racking forces on containers and their lashings. Although there are no financial gains to be made by the shipper who declares less than actual weight, the industry acknowledges that overweight containers are a problem. However, as yet this has not justified a requirement for compulsory weighing of containers prior to loading."[3]

---

[3] See U.K. Marine Accident Investigation Branch (2006, p. 31).

The incidents noted above did indeed prompt a multiyear study of the problem by the International Maritime Organization (IMO). In response, the IMO decided that the onus for accurate weighing would be on the shipper. The Safety of Life at Sea (SOLAS) convention as of July 1, 2016 requires container weight verification be provided to the carrier on a document and this will be a condition for vessel loading. Shippers will have two options in order to comply. The first is to drive the container over a weighbridge and subtract the weight of the truck, chassis, and fuel. The second option is to weigh each cargo item (including packaging, pallets, and securing material). The sum of these items' weights is to be added to the weight of the empty container. If a container is found to be in noncompliance, the vessel may leave with container held at the port for either return to the shipper or to the proper authorities. Unlike CSI, the carrier is the intermediary simply passing the document along. It does not bear an onus to insure that the shippers comply. Of course, it is the carrier who takes on the operational risk against safety when weights are not accurate.

THE SECURITY-EFFICIENCY TRADE-OFF

There is no doubt that political choices, as well as purely economic ones, influence decisions to regulate or to de-regulate. In the wake of 9/11, some politicians called for the physical inspection of all containers imported to the United States. Of course, such a policy would have reduced trade to a trickle as bottlenecks formed at ports of entry, just-in-time purchasing would become impossible, and global supply chains would retreat more and more within domestic borders. In compromise post-9/11 regulations intended to create new rules which established a form of certainty for shippers and carriers; and with certainty comes industry security. The choice of degrees of security over efficiency is a value judgment, of course, and the desired mix of these two items for an industry and for a society is a decision to be made at the political level. As political moods change, so will the desired mix.

Regulation produces degrees of supply chain security. In this regard, regulation is presented in the form of a typical production function in the left-hand panel of Fig. 7.1.

With a low degree of regulation, increasing marginal returns occur in the form of establishing the rules for reliable market transactions: property rights, contract enforcement, information dissemination, etc.

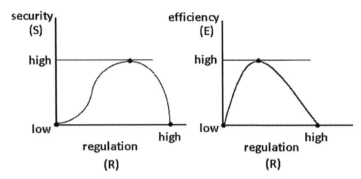

FIGURE 7.1
Security and efficiency with regulation.

Of course, as regulation increases further, diminishing marginal returns set in due to extra resources being devoted to compliance. In this situation, the degree of security is rising because evermore aspects of business are being subject to specific rules or protocols. Each firm, however, is trading off resources used for expansion and for research and development in order to spend money and time on lawyers, internal auditors, and safety inspectors all in the name of compliance. Each unit of security is harder to come by from an extra unit of regulation. Finally, negative marginal returns set in once regulation goes beyond that level necessary to maximize the security.[4] It is here that bureaucratic red-tape in the form of overregulation begins to stifle business activity. It may also be the case that regulatory fines become so high that firms respond by scaling back operations out of fear of making a mistake in procedure or facing a court challenge. The level of security falls because excessive regulation has now replaced market-uncertainty with compliance-uncertainty, which is, nonetheless, a form of business uncertainty that is detrimental to the operation of firms. Supply chain regulations set in the spirit of partnership cannot create compliance-uncertainty; but the danger is that the police function can lurch into the area of negative returns if it is not well thought out.

The right-hand panel in Fig. 7.1 shows the relationship between regulation and efficiency. The asymmetric shape shows that efficiency gains

---

[4] The maximum security shown in the figure is really that which mitigates risk given current technology, the ability to gather and interpret Big Data, etc. It should not be considered a bliss point on its own.

from regulation arise quickly. Before the region of negative marginal returns from excessive regulation sets in, it is preceded by a region of diminishing marginal returns. In that region each unit of regulation leads to greater efficiency but does so in decreasing marginal units. Why? Because each unit of regulation is less important to the achievement of market efficiency than is the previous, meaning each unit adds less and less to the market's ability to adjust to economic changes; that is, to be efficient. This is akin to the process of diminishing marginal utility for a good or service on the part of a consumer. When negative marginal returns set in, regulation is hampering the market: (1) in its ability to allocate resources to the production of those goods and services desired by consumers; and (2) to allow firms to produce them using a least-cost combination of resources.[5]

It should be noted that the level of regulation that brings about the maximum point on the efficiency curve is likely to be far lower than the level that maximizes security. This is the heart of the concept of the S-E trade-off. Fig. 7.1 showed regulation begetting levels of security and efficiency; therefore, a given level of regulation will beget an S-E pair. Fig. 7.2 sets out an appropriate trade-off curve.

The region bounded by R and R′ in quadrant I sets out the trade-off whereby one service attribute may only be gained at the expense of the other. Quadrant II is the security function from Fig. 7.1 drawn, instead, from right to left. Quadrant IV is the efficiency function from Fig. 7.1 rotated clockwise by 90 degrees. Finally, quadrant III shows a 45-degree line indicating the range of possible levels of regulation with the maximum being labeled as R_{max}. This line is used to capture all S-E pairs brought about by a common level of regulation. Consider, for example, point R′ on the 45-degree line in quadrant III. A vertical line drawn from it to the security curve in quadrant II gives the level of security that accompanies that degree of regulation. Drawing a horizontal line from R′ to the efficiency curve in quadrant IV gives a level of efficiency. These two levels of security and efficiency are traced from their respective

---

[5] During the period of interstate regulation of for-hire trucking in the United States, the Interstate Commerce Commission (ICC) had been known to deny trucking firms the right to carry direct backhaul freight thus preventing least-cost routing to some firms. Any route deviations necessary to fill backhauls added extra costs to the round trip. Furthermore, a filling of empty backhauls can, under certain conditions, serve to lower front haul freight rates as well. See Felton (1981) and Prokop (1998).

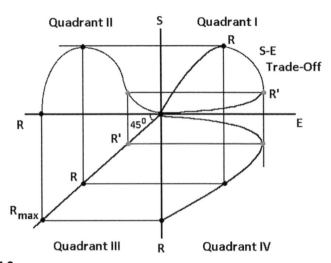

FIGURE 7.2
The security-efficiency trade-off.

quadrants over to quadrant I to the point labeled R′ as well. In this fashion, all points on the trade-off may be established.

Moving from R to R′ along the trade-off indicates a falling level of regulation—de-regulation in other words. Furthermore, the concave function that arises is indicative of the increasing opportunity cost of efficiency in terms of security foregone. This is because R and R′ represent the maxima of security and efficiency, respectively, brought about by specific levels of regulation. For a given increment of de-regulation (regulation), gains to efficiency (security) occur at a decreasing rate while security (efficiency) is foregone at an increasing rate as may be seen in Fig. 7.2.

A bliss point may be achieved through the addition of a social welfare function (i.e., an indifference curve for the whole society) to quadrant I in Fig. 7.2. With typically convex social preferences assumed, the bliss point would occur along the concave section of the trade-off. It is interesting to note that, by construction, society would then be operating at a level of efficiency that provided negative marginal returns to regulation. While operating at negative returns is uncommon in microeconomic theory, it must be borne in mind that the S-E trade-off has two moving parts, meaning that efficiency cannot be looked at in isolation. This is not a short-run model where efficiency levels may be chosen with security levels held constant and vice versa. Indeed, changes in regulatory levels are properly accounted for as a long run phenomenon of government activity.

It is also the case in Fig. 7.2 that the S-E trade-off backward-bends to the origin from both R and R'. In that sense a trade-off no longer exists. Increasing levels of regulation up to R' provide increases in both security and efficiency. In this case, an increase in regulation would definitely increase social welfare. The region along the curve between R' and the origin indicates an economy experiencing destructive competition, the absence of formal markets, or a lack of social order. By contrast, increasing levels of regulation beyond R bring about a decline in both security and efficiency leading to a lower level of social welfare. The region along the curve between R and the origin indicates efficiency losses in the marketplace combined with compliance-uncertainty among the regulated businesses. This area would be representative of too much policing and not enough partnering by government when it came to security programs. Regulatory rents are in a state of flux among incumbent firms and potential entrants thus leading to rent seeking expenditures among them.[6] Possibilities for rent seeking can occur in two fashions:

- In international trade: CBP, via the Advisory Committee on Commercial Operations (COAC), makes it harder for smaller shippers and carriers to comply with C-TPAT, FAST, etc., because of lack of clarity or subjectivity in the vetting of security plans. Also, ACE's "single window" does not occur in the near future but uncertainty of clearance at points of entry remains. These extra costs bias the system in favor of large shippers and carriers who can better manage the uncertainty.
- In emergency management: FEMA does not offer clear guidelines for prospective vendors and, in effect, confers incumbent power on those vendors with previously established relationships. This serves to limit competition and reduces incentives for cost control before or even during emergency situations.

---

[6] In a rent seeking environment the political outcome is definitely in a state of flux but the outcome itself is not random; rather, it is purchasable. In effect, the political game is not biased at the outset, meaning that an industry may find itself on this part of the S-E trade-off due to a rational transaction between government and industry to the exclusion of the broader interests of the electorate. If regulation is not used by government as a means to provide the two service attributes for the purpose of social welfare maximization—but rather to increase its own receipt of largess—the effect of rent seeking would be to restrict the industry to excessive regulation where negative returns to both security and efficiency are occurring.

Two points for elaboration regarding Fig. 7.2 are: (1) peak-efficiency (given by R′) is achieved before peak-security; and (2) the greater the spread between R′ and R, the greater will be the concave range of the S-E trade-off. With respect to point (1) it would appear realistic to set this as a requirement given that regulation always provides diminishing returns to efficiency while it starts out by providing increasing returns to security. A relatively low level of regulation is necessary to establish property rights for market efficiency while a relatively higher level is necessary to prevent compliance-uncertainty across established firms.[7] With respect to point (2) the spread will increase if the range of negative marginal returns in the security (efficiency) curve became less (more) pronounced up to the level R_{max}.[8] For the efficiency curve, this would imply that relatively less regulation is needed in order to bring about effective market conditions.

Is a positive-sum outcome a possibility? Yes, as noted in footnote 4, the security curve's maximum could increase under certain circumstances. In this case there is a potential for a higher level of security for a given level of efficiency. In Fig. 7.2, the R to R′ portion of the curve would expand outward potentially improving social welfare.

THE ECONOMICS OF SUPPLY CHAIN SECURITY REGULATION

The two attributes of regulation—security and efficiency—would be affected in a variety of ways depending upon the type of regulatory program the government puts in place. Security may be measured in

---

[7] If the peak for security occurred before that of efficiency, points R and R′ would exchange places on the trade-off curve but not on the 45-degree line in Fig. 7.2. Increasing levels of regulation would trace out the S-E trade-off in a clockwise movement. In Fig. 7.2, it is traced out in a counterclockwise movement. If it were the case that both peaks occurred at the same level of regulation, there would be no trade-off; in fact, the S-E "trade-off" would become a ray extending to the bliss point as traced out from the two peaks.

[8] R_{max} is convenient to use in that, by having both efficiency and security drop to zero at that common level of regulation, it defines the origin as the point where the two backward-bending portions of the S-E trade-off meet. For example, if the range of negative returns along the security curve extended beyond R_{max}, efficiency would be zero while positive levels of security would still exist. The backward-bending portion of the S-E trade-off from R would intersect the security axis instead of the origin.

terms of an absence of market volatility. In the context of supply chain security, volatility can run the gamut from shippers/carriers/brokers not knowing the proper data to send, as well as when and where to send it. On the other side of the security equation, volatility occurs if CBP does not know how to screen the incoming data in order to make the screening and inspecting processes systematic for shippers/carriers/brokers as opposed to appearing seemingly random. Each side does not fully comprehend the processes of the other. From an enterprise resource planning (ERP) perspective the processes are not mapped properly. From a game theoretic perspective the two sides are not coordinating properly. They are achieving the inferior Nash equilibrium in the coordination game of Fig. 2.2. Only improved coordination would achieve the Pareto-optimal result.

Efficiency may be measured in terms of the removal of excess burdens (i.e., losses in economic welfare) in the markets or, more normatively, measured as the degree of establishing the conditions necessary to conduct business (i.e., importing and exporting). In the context of supply chain security, problems can run the gamut from the extreme of closing points of entry due to a terrorist event to having to physically inspect cargo due to a problem with the incoming data from shippers/carriers/brokers or from problems with the CBP's screening technology indicating false positives for a security breach. If each side is living up to the other's expectations then the partnership, as a form of market transaction, is efficient. Each side believes the other is acting in good faith. From an ERP perspective, the process mapping is facilitating trade and not disrupting it; and it is allowing program benefits to be realized (i.e., single window data entry and expedited entry via C-TPAT and FAST). From a game theoretic perspective, the two sides have established trust in partnership. They are achieving the Pareto-optimal outcome of the prisoners' dilemma game of Fig. 2.1.

Where would businesses end up on the S-E curve? Simply specifying an equilibrium as given by the S-E trade-off and some convex social welfare function would leave out much of the developments in the economics of regulation that have been made over the last few decades. Consider the most relevant theories of regulation applied to supply chains in the light of the S-E trade-off.

Regulation may be looked upon as a means to correct market failures (e.g., negative externalities caused by firms not vetting their upstream vendors' security plans). Thus, regulation would be for the public interest. When applied in the market failure context, efficiency gains occur; the effect on security occurs when it is applied in the equity context.[9] This theory is normative in that it indicates the context under which regulation *should* occur in order to serve the goal of social welfare maximization. As such, this theory of regulation applies to the portion of the S-E trade-off from the origin through R' and up to R. There is no explanation for how the economy may operate where both service attributes are declining; nor is there an explanation as to why regulation occurs. There is also a paternalistic view of regulators put forward here in that the regulator will be working to correct a situation that the market cannot or will not.[10]

Alternatively, the "capture theory" asserts that regulation is used for the promotion of a firm's profits and not social welfare. Depending upon the efficacy of firms in playing the political game necessary to influence regulatory decisions, the assertion is that regulators serve the regulated firms based upon the latter's specialized knowledge of how their industry works. This means that regulation will surpass peak-efficiency on the S-E trade-off and, at best, operate where diminishing returns to security are occurring. Regulations are always pro-producer under this theory. Before de-regulation in the for-hire trucking industry, for example, entry was controlled and rates were set above cost as noted in Prokop (1998). The capture theory appears to indicate a position along the S-E trade-off anywhere above R', through R, and to the origin. But while the capture theory is not normative, it cannot explain de-regulation—a clockwise move down the S-E trade-off—because the shippers/carriers/brokers

---

[9] This type of public-interest regulation has come to be known as "normative analysis as positive theory." See Viscusi et al. (1998) and Joskow and Noll (1981). It was due to the lack of empirical support for this point of view that other theories of regulation developed that attempted to make the actions of regulators endogenous.

[10] In terms of viewing regulation as a service arising in a principal-agent context, the regulator (as agent) for the shippers/carriers/brokers (as principal) is, in effect, applying the proper "medicine" to business prospects in the marketplace. In this way, the relationship between principal and agent is one akin to a patient and his physician. The patient is submitting to the perceived wisdom of the physician.

represent the only interest group in the analysis.[11] The capture theory does not seem to apply very well to post-9/11 supply chain security because the government can take on the role of police which is not usually in the interest of industries preferring to self-regulate.

Another more subtle model of influence and lobbying was proposed by Becker (1983). In this case, interest group competition is made more explicit. Interest groups compete via political pressure for a fixed level of regulatory wealth. The aggregate level of pressure is assumed fixed, and the group or groups that receive a transfer obtain it at the expense of the remaining group or groups which are to be taxed in order to finance it. But the aggregate transfer is less than aggregate tax due to the transaction cost of redistribution. In this sense, regulation arising out of interest group competition leads to an aggregate welfare loss. Since the influence of any one group depends on its own potential pressure, as well as that of its rivals, a prisoners' dilemma game arises which leads to a political (Nash) equilibrium.[12] The application of pressure is costly but, on the other hand, a drop in one group's pressure increases the relative pressure of the others thereby creating a transfer in wealth across the groups. With two or more groups attempting to each increase its own absolute influence on regulators, its relative influence will nonetheless remain constant in equilibrium. Because of noncooperation, all groups spend more to increase their transfer or reduce their tax than if they had cooperated. Therefore, the political equilibrium is not Pareto-optimal.

The subtleties in interest group competition can be seen in the structure of the COAC subcommittees discussed in Chapter 6. For example, there would be conflicting interests in trade facilitation through the removal of trade barriers versus security through setting up non-tariff trade barriers. The global supply chain subcommittee and the trade enforcement and revenue collection subcommittee represent these different interests. On the other hand, COAC sets up a mechanism for industry and

---

[11] In the principal-agent context of service provision, the capture theory relegates the regulator to blindly pursuing the interests of the industry he is regulating. In this context, the principal is a *client* of the regulator who is performing well-defined activities for his client's benefit. Specifically, this involves a level of security not necessarily desired in terms of social welfare maximization.

[12] With the political equilibrium being a Nash equilibrium, it is therefore required that each group simultaneously decide upon the level of pressure to apply.

government partnership and may improve the political outcome relative to a simple top-down approach.

Becker's model has an affinity with the public-interest model of regulation. The applicable range of the S-E trade-off would, therefore, be the R to R' region, where social welfare may be maximized. Consider this affinity more carefully. Interest group A would receive a lower transfer if the marginal welfare loss (w) of a tax on interest group B increased. Group A would lower its pressure expenditures while group B would increase its pressure in order to avoid such a tax levy. The transfer to group A is therefore reduced, meaning that regulation (by way of transfers) decreases in the industry less prone to welfare losses. Conversely, regulatory policies are more likely to increase social welfare. Furthermore, any presence of market failures means that w is low (or possibly even negative) and the group so-affected has an incentive to increase pressure. Therefore, regulation *tends* toward the optimal combination of S-E (if both service attributes are taken to be merit goods that the regulator knows how to provide in conjunction with the social welfare function).[13]

ISSUES AND PROBLEMS GOING FORWARD

The S-E trade-off may be seen as a useful way of assessing the viability of economic regulation. Regulation provides two distinct market conditions: supply chain security and market efficiency. Regulation is certainly a principal-agent activity.

Post 9/11 security programs are noted for their reliance on government partnerships with shippers and carriers, as opposed to the government setting up a bottleneck of rules at a point of entry. Proactive groups are looking to form alliances and suggest standards for government to consider. Two important groups in this regard are the intergovernmental World Customs Organization and the Global Air Cargo Advisory Group which is made up

---

[13] The Becker model, by its own assumptions, may not always follow the results of the paternalistically structured public-interest model. The application of pressure is, by assumption, subject to diminishing marginal returns for each interest group. Differences in the degree of these marginal returns across different groups are just as much a determinant of the pressure applied as are the welfare losses of taxation. A more efficiently organized interest group faces less pronounced diminishing returns and can therefore be expected to provide more pressure.

of special interest groups dealing with air carriers. In sum, supply chain security is a shared responsibility while trade is a shared benefit. Given the nature of the supply chain security programs that are in place and the political issues involved in balancing security with efficiency, the question is how to measure success in these programs. This is the topic of Chapter 8.

Bibliography

Becker, G.S., 1983. A theory of competition among pressure groups for political influence. The Quarterly Journal of Economics 98, 371–400.

Donovan, A., Bonney, J., 2006. The Box that Changed the World. Commonwealth Business Media, East Windsor, NJ.

Felton, J.R., 1981. Impact of ICC rate regulation upon truck back hauls. Journal of Transport Economics and Policy 15 (3), 253–267.

Joskow, P.L., Noll, R.G., 1981. Regulation in theory and practice: an overview. In: Fromm, G. (Ed.), Studies in Public Regulation. MIT Press, Cambridge.

Levinson, M., 2006. The Box: How the Shipping Container Made the World Smaller and the World Economy Bigger. Princeton University Press, Princeton, NJ.

Prentice, B.E., 1994. The stability/efficiency regulatory trade-off: policy implications for for-hire trucking. In: Canadian Transportation Research Forum: Proceedings, pp. 494–507.

Prokop, D., Prentice, B.E., 2009. Regulation as customer service: a cyclical view. In: Ross, A.D. (Ed.), Supply Chain Management in a Global Economy. Lombard, II: Council of Supply Chain Management Professionals. Proceedings of the 2009 CSCMP Supply Chain Management Educators' Conference.

Prokop, D., 1998. The Canada-U.S. Transborder Trucking Industry: Regulation, Competitiveness and Cabotage Issues (Ph.D. Dissertation). University of Manitoba.

U.K. Marine Accident Investigation Branch, 2006. Report on the Investigation of the Loss of Containers Overboard from P&O Nedlloyd Genoa. Report 20/2006.

Viscusi, W.K., Vernon, J.M., Harrington Jr., J.E., 1998. Economics of Regulation and Antitrust, second ed. MIT Press, Cambridge.

CHAPTER 8

Mitigating Risk and Measuring Success

CONTENTS

COMPLEXITY AND RISK IN THE SUPPLY CHAIN

Complexity

Supply chain management is complex because of its structure; that is, the variety of players and the distances involved in a globalized environment. It is also challenging because of its inherent risk. As to structure one just has to start with the end customer and work upstream back to the retailer, the wholesaler, the manufacturer, and the primary product. Of course, this is too simplified since the retailer deals with several wholesalers who, in turn, deal with several manufacturers, etc. Every one of these relationships is held together by a legal contract. An upstream supplier may very well supply to two or more customers who are in competition with each other; but it is the contracts between the buyer and the seller which are unique and provide a potential competitive advantage of one supply chain versus another. Of course, in-between all these relationships are the transportation carriers and intermediaries which help facilitate domestic and international trade.

151

Global Supply Chain Security and Management. http://dx.doi.org/10.1016/B978-0-12-800748-8.00008-X

Finally, the government (in the form of various regulatory agencies) helps or hinders these flows of trade depending on what the current policy is. All of this takes place within an environment or state of nature which none of these parties can control. This is a primordial source of risk to all supply chain plans.

Of course, risk exists in human interactions as well because plans need not work smoothly. The end customer's desire to buy or not to buy is governed by tastes and preferences and the perceived value of the item relative to the price to be paid. In other words, a lot of psychology governs the desire to buy. On the other hand, sellers are governed more so by the laws of physics; that is, what level of inputs are available and what level of productivity is possible in order to meet all or some of the demand for goods. In competitive markets there is an incentive for suppliers to control costs lest they be priced out of the market. Striving for cost control, combined with the Internet's ability to help quickly locate possible vendors over vast spaces, has led to the extension of supply chains over longer distances. On the one hand this mitigates the risk of being tied to an underperforming vendor; but on the other hand the longer the supply chain the higher the risk from the environment grows.

Complexity, and the realization that everything cannot be controlled, means that a certain amount of risk in the supply chain must be tolerated. Supply chain risk may be defined as the likelihood that an event with negative consequences will occur. The impact of the event, from a single organization's point of view, depends on where it resides along the supply chain relative to where the event occurred. The shock wave dissipates to some degree as it leaves the epicentre; but the degree depends on the situation. Supply chain risks may be broadly characterized in the following way:

- External and either localized or end-to-end: for example, a power outage at one plant would be local while a terrorist event (even if directed at one plant) can have end-to-end effects. Dependencies along the supply chain determine how the shock wave will ripple along.
- Suppliers: an upstream risk of a slowdown in the receipt of inputs in the production process.
- Distribution: a downstream risk which slows down the process of getting outputs to market. For example, a carrier may not have the capacity to ship.

- Internal: a breakdown in the production process. For example, a labor strike could shut down a plant. Quality control could be a problem.

Of course, the government has an impact as well on these supply chain risks. Political uncertainty, currency fluctuations, and changes in the tax regime can increase external risk along the supply chain. Specific taxes, trade tariffs, and non-tariff barriers (such as compliance with supply chain security programs as covered in Chapter 4) can affect the level of risk in dealing with upstream supply and downstream distribution. Internal risk may increase due to wage and price controls, taxes, labor laws, etc. The overall question is: Can some of these risks be mitigated? If so, how?

Risk Mitigation

Supply chain management, by necessity, encompasses risk management. If a manager acknowledges that outcomes are subject to variability which can lead to disruption then there is risk. These outcomes are often unplanned and unanticipated. Part of the manager's job would be to help mitigate this risk.

Chapter 3 outlined the range of threats the supply chain faces today. Chapter 5 expanded the discussion to cyberspace. While it is important to take stock of threats, the focus must be on the areas at risk. As Harrison (2010) notes: "Threat is an exploitable vulnerability. When one examines the range of potential targets the modern supply chain offers to a prospective criminal and terrorist, the threat appears enormous and unmanageable."[1] Following this line of reasoning the way to measure risk is to assign a probability to a particular threat being exploited. If a dollar value of the consequences of the crime is estimated, then an expected cost can be assigned. The expected cost should be balanced against the actual cost of mitigation. Naturally, the organization tasked with mitigation would like to see mitigation cost below the expected cost of the crime. A simple key performance indicator (KPI) is the ratio of the cost of mitigation to the expected cost of a successful crime; and this KPI needs to be less than one. Another

---

[1] Harrison (2010, p. 54). Of course, one way to mitigate risk is through insurance. This does not eliminate risk but merely transfers it to the insurance company and its client pool. For a review of various types of insurance applicable to supply chains, see Fischer et al. (2013, pp. 153–161).

useful KPI would be "time to recover" from an event. This highlights the opportunity cost of not investing in risk mitigation. During recovery there is a loss in revenue and, potentially, loss of life and brand reputation. The difference in the loss during the recovery time with particular forms of mitigation versus without helps to assess their worth to the organization.

Another simple method is to categorize the probability of a supply chain disruption as either high or low and the magnitude of the disruption as either high or low. Possible events would be assigned to one of four cells in the following matrix[2]:

Decisions makers need to focus their attention on the high vulnerability cell. Events in this cell have the highest expected cost to the organization and its supply chain. From a crime/terrorism perspective the organization is target-rich and provides an enticing opportunity (i.e., high probability of success in the eyes of the criminal/terrorist). From a natural disaster perspective the organization would be located in an area prone to earthquakes, tornadoes, hurricanes, etc. and these events should be listed in that cell.

Vulnerability is the mirror image of resiliency; therefore, the low vulnerability cell would include disruptions that the organization can deal with quite effectively through minor adjustments or repairs. From a crime perspective this includes petty theft and vandalism due to, say, forgetting to lock a door or window. From a natural disaster perspective this includes thunderstorm damage, minor flooding, etc. The normal challenges cell represents expected operational challenges that should also be easy to deal with. The best example of these includes disruption to infrastructure (e.g., port congestion). As long as other transportation options are available, the resilient organization should have alternative plans in place to detour incoming and outgoing shipments. Finally, the speculative planning cell includes events which are rare but highly impactful (e.g., known unknowns and black swans as discussed in Chapter 2). From a crime/terrorism perspective this would include a criminal or terrorist breaching a "state-of-the-art" security system. From a natural disaster perspective this would include unseasonable severe weather (see Fig. 8.1).

---

[2] This matrix builds on Sheffi and Rice (2005, p. 43).

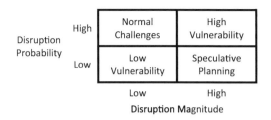

		Low Disruption Magnitude	High Disruption Magnitude
Disruption Probability	High	Normal Challenges	High Vulnerability
	Low	Low Vulnerability	Speculative Planning

FIGURE 8.1
Vulnerability matrix.

Given the nature of the organization and its supply chain it does not have much control over whether or not an event's disruption magnitude is categorized as low or high. For example, a terrorist event is likely to always be highly disruptive while petty theft will not. The organization does, however, have control to some degree over the probability of the disruption of some events. For example, deploying "state-of-the-art" security systems can move the incidence of terrorism from the high vulnerability cell to the speculative planning cell. In this case, risk mitigation appears to have worked; but the organization has to make sure to plan for the unlikely event that the security system fails. This means building in back-up systems, redundancies, alternative sources and facilities, etc. From a natural disaster perspective a state-of-the-art facility in an earthquake zone may mitigate the damage of low magnitude earthquakes but may not mitigate the worst unless well thought out plans from upstream and downstream supply partners in and outside the impacted area are in place.

Engagement

Organizations should put in place cross-functional teams of experts with representatives as far along the supply chain as they can engage. Not only does this assist with the planning process, it also assists with regaining control of the situation when a supply chain disruption occurs.[3] Fig. 8.2 gives an impression of the activities along a timeline.

---

[3] See Sharma and Vasant (2015, p. 10). The teams and the security plan presuppose cooperation among suppliers, manufacturers, distributors, carriers, intermediaries, and the government. The plan should cover the protection of all supply chain assets against disruptions and prevention of contraband. This can be handled through designing rules and using appropriate technology in order to prevent disruptions and, if unsuccessful, resume operations within minimal loss.

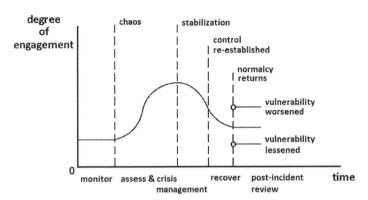

FIGURE 8.2
Disruption management timeline.

The organization starts at time zero with a security plan in place and a degree of engagement with a cross-functional team which it believes is commensurate with the possible disruption. The disruption may or may not occur; but if it does the activities in Fig. 8.2 show a process which needs to be accounted for in the security plan.

Assume that after some time of monitoring the situation a disruption occurs. The opening moments are one of chaos which forces the team to accelerate the level of engagement as it tries to assess the situation. Crisis management, as outlined in the plan, is deployed after the proper assessment has been made. If it is appropriate and mitigating then an inflection point is reached and the engagement begins to peak as the disruption is stabilized. Once the team feels that control has been reestablished the process of recovery begins and engagement levels off. Once normal working conditions return, it is very important for the team to conduct a comprehensive post-incident review in order to learn from the disruption and how it was responded to.

Naturally, the timeline for these activities is dependent on the type of disruption that has occurred. If it is the case that the disruption is impactful to the point where government intervention is necessary (i.e., it is a disaster) the timeline for chaos and stabilization would expand since multiple supply chains would be affected and the government's response has to be prioritized. Recall Fig. 4.1 and how the government's first task may be to establish the strategic goals of law and order, public safety, and security. The tactical provision of water, food, shelter, medicine, and energy (power) will take precedence over the restoration of any private supply chains.

Crisis management as depicted in Fig. 8.2 does have a common factor whether the disruption is to a private supply chain or to a wider public area. Assessing the crisis can be messy and chaotic. As such, this shows the effectiveness (as opposed to the efficiency) of the crisis managers. Effectiveness involves getting a handle on the disruption/disaster as fast as possible with as much resources as possible. This "all hands on deck" approach involves such things as: locking down the impacted facility, guarding critical infrastructure, putting out fires, and, above all else, saving lives. This is not a process based on market signals and negotiating. Efficiency, which does require responding to market signals, should only enter into the calculations once normalcy has returned.

The post-incident review may involve improved training, adjusting responsibilities among supply chain partners, and possibly changing partners. The review should develop questions asked and lessons learned. These should be worked into the updated security plan. Once this is done it may or may not be business as usual. There are two possibilities that may arise. The learning and adaptation to the past disruption may indicate that a structural change has occurred in the state of nature for this supply chain. The vulnerability inherent in the organization may be perceived to be permanently higher or lower and requires the respective degree of engagement. The disruption, in other words, would be a game-changer.

Certainly the degree of engagement ought to be positively correlated with the perceived probability of the disruption. How can the probability of disruption be measured? Certainly it is a subjective process; however, the process is more than just a guess. It is a matter of one gathering intelligence and being able to think like one's adversary. Harrison (2010) suggests three components of useful intelligence worth gathering about criminals:

1. What is their intention?
2. What are their capabilities in carrying out their intentions? and
3. What is the nature of the environment they operate in?[4]

In other words, what do they wish to do, how would they do it, and what is the origin and direction of the action to take place? This is very similar to the law enforcement concepts of motive, means, and opportunity. Intelligence is a crucial part of assessing the state of nature.

---

[4] Harrison (2010, p. 55).

Intelligence Gathering

The process of turning data and facts into knowledge, understanding, and wisdom (as discussed in Chapter 5) certainly applies to intelligence gathering. The process of gathering, analyzing, and interpreting data/facts on other nations and hostile forces (whether foreign or domestic) occupies most of what the government's intelligence community works on. The idea is to start with the data and the facts and begin to explore the unknown.

The Joint Counterterrorism Assessment Team (JCAT) identifies six types of intelligence sources[5]:

- Geospatial: representation of physical features on earth and geographic points of reference.
- Human: includes espionage and interrogation techniques. It may also be gathered through standard international diplomacy.
- Imagery: representation of objects by electronic means, film, or other media.
- Measurement and Signature: data analysis leading to distinct characteristics of a target. It is gathered through a variety of means. These include acoustic, chemical, optical, radar, radio frequency, seismic, etc.
- Open-Source: information available to the general public. This includes mass media, government and business reports and speeches, and public observations (e.g., amateur photography, radio monitors, Internet chatter, etc.).
- Signals: electronic or verbal communication and transmission in a standard or coded language.

The data and the facts are processed into "finished intelligence" through the techniques of translation, decryption, and interpretation. It comes in five forms:

- Current Intelligence: relates to day-to-day threats and their perceived short-term impacts.
- Estimative Intelligence: relates to assessment of long-term threats and alternative scenarios. Prominent examples are the National Intelligence Estimates produced by the National Intelligence Council and relate to the United States' national security.

---

[5] JCAT (2015, pp. 20, 21, and 48).

- Warning Intelligence: relates to highly probable threats, or to those with low probability but high levels of disruption.
- Research Intelligence: leverages current and estimative intelligence to produce a more in-depth study of a threat.
- Scientific and Technical Intelligence: relates to the study of threats posed by weaponry and their technical support systems.

Finished intelligence is supposed to provide the decision maker with situational awareness involving: insights, assessments, and warnings. But the intelligence gathering process has two caveats: (1) it is a forecast and, as such, cannot remove all uncertainty when dealing with complexity; and (2) it cannot violate the law.[6]

The US intelligence community is a very large and diverse part of the federal government's bureaucracy. It includes various military, law enforcement, and civilian organizations:

- Air Force Intelligence
- Army Intelligence
- Central Intelligence Agency
- Coast Guard
- Defense Intelligence Agency
- Department of Energy Office of Intelligence
- Department of Homeland Security, Office of Intelligence and Analysis
- Department of State, Bureau of Intelligence and Research
- Department of the Treasury, Treasury Office of Terrorism and Financial Intelligence
- Office of the Director of National Intelligence
- Drug Enforcement Administration, Office of National Security Intelligence
- Federal Bureau of Investigation, National Security Branch
- Marine Corps Intelligence
- National Geospatial-Intelligence Agency
- National Reconnaissance Office
- National Security Agency
- Naval Intelligence

---

[6] Applicable laws include: the *National Security Act* (1947), the *Foreign Intelligence Surveillance Act* (1978), the *Intelligence Reform and Terrorism Prevention Act* (IRTPA; 2004), the *Privacy Act* (1974), the *Detainee Treatment Act* (2005), the *Homeland Security Act* (2002), and the *Military Commissions Act* (2006).

As intelligence gathering organizations they have in common not the search for truth but, rather, the search for approximations to the truth. This is because they are dealing with perceptions in the face of complexity. The skill which good intelligence officers hone is to abandon the fear of being wrong, to let the data/facts speak, and to constantly update their information and their estimates.

MEASURING SUCCESS

The Basics

Supply chain security is a goal of both business and government. It is a shared responsibility that businesses are beginning to appreciate as a source of competitive advantage.[7] Government appreciates supply chain security, in part, from its willingness to explore public-private partnerships. Supply chain security planning and compliance are public goods.[8] Chapter 2 looked at coordination of private businesses with the government in a game-theoretic context. However, the nature of partnership among the private businesses themselves can be illustrated using game theory as well.

In considering supply chain security programs such as C-TPAT and CSI shippers and carriers each apply for membership and interface with CBP. Furthermore, the shippers and for-hire carriers are dependent on each other in terms of their respective efforts of compliance. In other words, the security chain is only as secure as its weakest link. In the game shown in Fig. 8.3 the shipper and carrier both independently choose one of two levels of security compliance: strong or weak. Security compliance takes the form of allocating the time and money toward the technology and protocols necessary to satisfy CBP. This two-party public good is represented by a "game of chicken." Both parties want a level of supply chain security even if they paid for it alone.[9] The risk in this arrangement is that if one chooses to be compliance-strong the

---

[7] For a review of a set of executive interviews on this matter, see Autry and Bobbitt (2008).

[8] Chapter 5 noted that the provision of cybersecurity, as a public good, has a tendency to be underprovided even though society would be better off with a fuller provision. This applies to supply chain security as well. Private businesses must invest in security while the general public, which may or may not deal with the businesses in question, benefit when their secure supply chains lower the policing costs of government, the chances of shutting down ports of entry, etc. For a more detailed discussion see Prokop (2004).

[9] For a discussion of public goods as a game of chicken see Mueller (1989, pp. 15–17).

Carrier Compliance

		Strong	Weak
Shipper Compliance	Strong	(25,25) (c,c)	(10,30) (p,f)
	Weak	(30,10) (f,p)	(0,0) (u,u)

Payoff Order: (Shipper, Carrier)
Where: f>c>p>u and 2c>(f+p)

FIGURE 8.3
Security programs as a game of chicken.

other could simply free-ride on that decision by choosing to be compliance-weak. This means that overall security is less than otherwise. In other words, the private sector has an incentive to underprovide supply chain security.

The payoffs are defined as: coordinated security (c); free-rider security (f); paternalistic security (p); and uncoordinated security (u). Also, the summed payoffs of coordinated security (2c) outweigh the summed payoffs when one of the two players decides to be a free rider. Thus, the supply chain is more secure when each player coordinates than when one chooses to free-ride.[10] Like the coordination game in Fig. 2.2, this game has two Nash equilibria: one partner chooses weak (and saves time and money by free-riding) while the other chooses strong (and has to pick up some of the slack with further investments of time and money). Compared to the Nash equilibria the choice of (strong, strong) is not a Pareto-improvement; however, the public good would be best served by both players choosing it. How can this result come about? The two players could bind themselves with a contract (as a credible commitment to adopt strong compliance). Recalling the discussion of ACE in Chapter 4, the government's security programs have affected the state of nature in two ways. First, both international shippers and

---

[10] Technically, the game of chicken still applies if (p) were 20. In this case $2c = (f+p)$. However, this would imply that if one partner plays for weak compliance the other's strong compliance makes up for the difference such that the aggregate result was as if both players were strong. In the real word it is not likely this could occur because of the specialized knowledge and experiences involved in transportation services versus those of a shipper's importing and exporting.

State/Local Agency

		Proactive	Reactive
	Proactive	(10,20)	(0,0)
FEMA			
	Reactive	(0,0)	(20,10)

Payoff Order: (FEMA, State/Local Agency)

FIGURE 8.4
Disaster relief programs as a battle of the sexes.

carriers will have to use ACE and provide the information requested. One cannot free-ride based on the compliance level of the other; both must, in effect, be compliance-strong in order to complete their parts in the process of international trade. Second, since shippers and carriers are vetted by CBP in order to join the C-TPAT and FAST programs there is an incentive for program members to self-select one another. In this way, there is less chance for one player to free-ride off the compliance level of another.

Another example of problems with partnership can be seen in the inter-governmental relations in disaster relief. Recall the discussion in Chapter 4 where FEMA and state/local authorities try to decide between proactive and reactive approaches to disaster relief. Unlike the coordination game in Fig. 2.2 the two players want to coordinate their efforts but each puts a higher value on different plans. Fig. 8.4 shows this "battle of the sexes" game.[11]

The game is structured with FEMA's preference for reactive planning and state/local agencies preferring proactive planning. However, both players see a lack of coordination as being the least valuable position. If this is the case, there will be two Nash equilibria of (proactive, proactive) and (reactive, reactive). Which equilibrium would more likely come about? There are three ways to decide. First, the players could negotiate a binding

---

[11] The game gets its name from the story of a married couple which wants to do things together but the husband most prefers a sports match while the wife prefers the ballet. In the pure coordination game in Fig. 2.2, both the husband and wife would feel the same way about sports and the ballet thus making the joint attendance at the most preferred event Pareto-optimal.

	State/Local Agency	
	Proactive	Reactive
FEMA Proactive	(30,30)	(0,0)
FEMA Reactive	(0,0)	(10,10)

Payoff Order: (FEMA, State/Local Agency)

FIGURE 8.5
Disaster relief programs as pure coordination.

contract which commits each to a particular planning approach. Second, one of the players simply makes a precommitment to one of the two plans. Most likely, the bulk of the political power rests with FEMA. If FEMA precommits to a reactive approach, state/local agencies may feel little choice but to coordinate their plans in that fashion. The game world settle on (20,10). Third, the decision is taken out of the two players' hands and turned over to a third party. In this case, a public-private partnership (PPP) like Alaska Partnership for Infrastructure Protection (as discussed in Chapter 4) in Alaska could work to get both players on the same page. The PPP would be made up of representatives of all levels of government in order to work to a consensus. In this way, Fig. 8.4 would morph into a pure coordination game which might look like Fig. 8.5.

In this example, the PPP feels that the appropriate plan should be proactive. There are still two Nash equilibria but the PPP informs the two players on how to value them. The point is that there is now a Nash equilibrium which is also Pareto optimal.[12] The PPP's hybrid approach is to be more subjective and choose the appropriate style given the situation at hand or what is likely to be at hand.

Sustained partnership is part of the formula for success. It will strengthen the supply chain. Shared intelligence is the raw material necessary to help turn data and facts into information, knowledge, etc. The supply chain needs to become resilient to disruption. This is a matter of effective security planning. Logistics needs to have more visibility. This is a matter of

---

[12] This transformation from one game to another can apply to other areas discussed so far. For example, supply chain partners with different levels of buy-in to ERP systems, tracking technologies, or choices of metrics are in a battle of the sexes game. Cross-functional teams which search for a compromise can change the problem to a pure coordination game.

real-time tracking, knowing routes, networks, having a chain of custody, etc. Investing in information technology and then preparing for the data avalanche is also necessary. This is a matter of overcoming the knowledge management problem of combining technology with employee buy-in.

Success can obviously be measured in terms of mitigating an incident. One way, given the types of intelligence noted above, is recognizing any heightened risk and trying to stop the threat from being carried out. This could be done through countering the criminal's capabilities after correctly assessing his intention. Another way is to change the environment in which the criminals thrive. This "draining the swamp" or "defeating the ideology" operation, as matter of government policy, can take longer but might yield more lasting results. There is a choice between a short-term goal of isolating a threat and a long-term goal of permanently defeating all sources of the threat. How far should mitigation be taken? Even if all crime could be eliminated is it worth it to eliminate all crime no matter how petty? If there is a "war of terror" and victory is possible, what does victory look like? In a business setting what does successful supply chain security look like?

Coming up with measurements is relatively easy. The harder part is coming up with ones which are meaningful. Before looking at some popular measures, the following items help clarify what an organization might want to see from a meaningful metric.

- Buy-in from all stakeholders. This is imperative for any cross-functional teams to be able to work properly and for partnerships and enterprise resource planning (ERP) systems to truly provide value. The teams and the ERP software users need to agree on the measures in order to speak a common language and work to a common goal.
- The measures need to be defined clearly and have a procedure in place as to how to measure them. Importantly, they need to link to the stated goals of the linked organizations. Ideally, the strategy should involve just a few easy-to-comprehend goals. In this way the "dashboard" to be monitored is not overwhelmed with measurements.

Finally, while too many measures are not helpful, so are too few. Ideally, the set of measures should be broad enough to cover data from the past and the present. Another measure should forecast the future. Measures could focus on variables such as: time, distance, volume, weight, cost, revenue, etc. If the measures comport with the goals, are reported in a scheduled and timely manner, and are in a level of detail tailored to the needs of the decision makers, the elements for successful measurement are in place.

Some Key Performance Indicators

There are several KPIs which supply chain security managers could use to monitor their operations. At business locations these could include:

- Response time to a breach in the physical perimeter or to a cyber firewall. This means the time for the system to react, the time for the authorities to arrive, and the time used to neutralize the incident. Using Fig. 8.2 as a guide this would be the time elapsed during the chaos and stabilization sections. Naturally, the shorter the time lapse the better.

In the context of transportation these could include:

- No breach to the cargo container, trailer, vessel hold, etc. This means no evidence of tampering with the doors, locks, or seals used to secure the cargo.
- No unauthorized stops or deviations from established routes.
- Chain of custody. There were no "black holes" in the transport and transfer of cargo downstream along the supply chain. A party stops tracking the cargo only when it is officially transferred to the next party. Naturally, electronic and real-time systems are preferred.

In the context of specific government programs these could include:

- C-TPAT: program compliance rate along the supply chain. This includes the percentage of partners meeting predefined CBP standards in their security plans; and the proportion of the partners receiving benefits at ports of entry. These benefits include time savings at FAST lanes.
- CSI ports: percent of on-time and accurate e-manifest submissions by shippers, carriers, and brokers. Proportion of containers pulled aside for further inspection before departure, on arrival, or both.

The time and monetary cost involved in collecting data makes it imperative that the chosen KPIs be aligned to specific objectives of the business and the collective supply chain partners. The KPI also needs to inform decision makers.

The Perfect Order Index and Security

The perfect order index is a particularly useful KPI. It is cross-functional and shows interdependencies across the supply chain rather than looking at each activity in isolation. The difference is akin to wanting to see the whole forest (and consider all possible threats within its boundaries) as opposed to examining a tree within it under the belief that it represents the entire forest. Such thinking avoids complexity and desires efficiency (i.e., precision) when what will be required is effectiveness (i.e., tamping down the disruption) in the face of the countless facets of the disruption that may hit the supply chain.[13]

The perfect order is one which is delivered:

1. On time at the right place (i.e., valuing time and place utility);
2. Complete (i.e., filled in full);
3. Damage-free[14]; and
4. With the correct invoice attached (i.e., paper or electronic and with accurate information)

The components are measured as ratios. For example, this involves the proportion of deliveries which were on time. The first and third components can be regarded as KPIs of the effectiveness of the supply chain security plan. The fourth component can be seen as a proxy for the proper handling of all documents required to complete an international transaction. Of course, as Chapter 4 discussed, customs clearance requires several documents (some of them to be sent well ahead of the shipment's arrival at the port of entry).

A business could separately track each of these four components. But since the perfect order, which the business wishes to provide, is an amalgam of these components, they are interdependent. The perfect

---

[13] Also, several disruptions can happen simultaneously, especially across a global supply chain. Since organizations have limited resources, there is no reason to assume that the effects would always be mutually exclusive. The result is that multiple simultaneous disruptions could magnify the path of a given disruption as mapped in Fig. 8.2. For a discussion of the mathematics involved in calculating combinations of multiple disruptions, see Bradley (2010, pp. 103–106).

[14] Damage-free, as a component of the perfect order, could also include theft-free since both involve product loss. The way to measure theft-free in this KPI would be the proportion of known theft attempts which were thwarted by intelligence or technology used by the business.

order index attempts to model this interdependence by multiplying all four components together. Assume all four components were operating independently at success rates of 90%, the interdependence of the perfect order index leads to a value of only $(0.90 \times 0.90 \times 0.90 \times 0.90) = 65.6\%$. It is also the case that if any one component changed by a certain percentage, *ceteris paribus*, the perfect order index would change by the same percentage. However, if more than one component changes in the same direction the perfect order index would change in that direction but to a magnified degree. Using supply chain security programs such as C-TPAT and CSI as examples, if a shipper or carrier uses these programs it is supposed to enjoy faster customs clearance. This should increase the success rate of component one. If shippers use carriers which are also members, this should increase the success rate of component three. Suppose these two components were to increase their success rates by 10%. This would turn a 0.90 success rate into 0.99. The perfect order index would then be $(0.99 \times 0.90 \times 0.99 \times 0.90) = 79.3\%$. This means that the perfect order index increased by 21%. In other words, the effect of supply chain security improvements can lead to magnified measurement effects because of supply chain interdependencies.

ISSUES AND PROBLEMS GOING FORWARD

As Fig. 8.2 showed, after the crisis is stabilized and normalcy returns, the organization and its supply chain partners press ahead with their businesses into a future that is far from certain. Taking the philosophy of logistics and supply chain management and coupling it with the need to secure the supply chain should lead one to wonder whether or not the process will be easier or harder in the future. One thing is certain, the process will be different. Some of the changes on the horizon will be discussed in the final chapter.

Bibliography

Autry, C.W., Bobbitt, L., 2008. Supply chain security orientation: conceptual development and a proposed framework. International Journal of Logistics Management 19 (1), 42–64.

Bradley, J.R., 2010. The complexity of assessing supply chain risk. In: Thomas, A.R. (Ed.). Thomas, A.R. (Ed.), Supply Chain Security: International Practices and Innovations in Moving Goods Safely and Efficiently, vol. 1. Praeger, Santa Barbara, CA, pp. 89–120.

Fischer, R.J., Halibozek, E.P., Walters, D.C., 2013. Introduction to Security, ninth ed. Butterworth-Heinemann, Waltham, MA.

Harrison, J., 2010. Supply chain security and international terrorism. In: Thomas, A.R. (Ed.). Thomas, A.R. (Ed.), Supply Chain Security: International Practices and Innovations in Moving Goods Safely and Efficiently, vol. 1. Praeger, Santa Barbara, CA, pp. 52–74.

JCAT, 2015. Intelligence Guide for First Responders. National Counterterrorism Center, Washington, DC. https://www.ise.gov/interagency-threat-assessment-and-coordination-group-itacg.

Mueller, D.C., 1989. Public Choice II. Cambridge University Press, Cambridge, UK.

Prokop, D., 2004. Smart and safe borders: the logistics of inbound cargo security. International Journal of Logistics Management 15 (2), 65–75.

Sharma, S.K., Vasant, B.S., 2015. Developing a framework for analyzing global supply chain security. The IUP Journal of Supply Chain Management 12 (3), 7–34.

Sheffi, Y., Rice Jr., J.B., 2005. A supply chain view of the resilient enterprise. MIT Sloan Management Review 47 (1), 40–48.

CHAPTER 9

The Future of Supply Chain Security

CONTENTS

NOWHERE TO HIDE

Criminals and Terrorists Under Surveillance

Consider the lone wolf. This is someone who is recruited to carry out an act of terrorism but has no criminal record or verifiable link to any criminal organization or terror group. Pairing such individuals with a weapon of mass destruction is a nightmare scenario. To guard against this requires the encroachment of the national security state even further into the lives of civilians. Instead of government agents listening to people talking in crowds or on street corners, they will monitor Internet chatter (i.e., postings on social networks, websites visited, etc.). Just like everyone gathered in the proverbial drawing room is

Global Supply Chain Security and Management. http://dx.doi.org/10.1016/B978-0-12-800748-8.00009-1

taken to be a suspect by the detective in crime novels and films, we can all be considered real-life suspects when trolling the Internet. Privacy will be further sacrificed since there is little expectation of privacy on the Internet.

As far as goods in storage or parked vehicles are concerned inexpensive mini-cameras can monitor any surrounding movements. As far as transportation, geo-fencing may be applied to every commercial vehicle. Of course, if the criminal is detected and takes flight, a drone could follow him based on his heat/motion signal. GPS coordinates could then be gathered in real-time. The data gathered by these processes may help speed law enforcement.

Access to secure areas could be enhanced through fingerprint, voice, and face recognition software. This could prevent inside jobs. Scanners can be used to detect the presence of people via the smartphones they are carrying. From there the scanner can identify the machine access control (MAC) address or Internet protocol (IP) address and, in effect, obtain the smartphone number of the intruder.

Businesses Under Surveillance

While criminals and terrorists are under increasing scrutiny, it is because of the reaction to the threat businesses feel. As previously discussed, more transportation and more items held in inventory are sources of vulnerability. Sharing more information online with supply chain partners and the government adds to the threat of cyberattack.

Cyberspace will increase to an astonishing degree over the next decade or so as the "Internet of Things" unfolds. While cybersecurity planners work to latch down computer and telecommunication networks, their task will be expanded to cover a host of other devices to be interconnected via wireless technology. These will include multiple billions of items including automobiles, home appliances, medical equipment, factory machines, etc. From a safety and security perspective, it is one thing if a personal computer is hacked and the hard drive crashes; it is quite another if it happens to an automobile in motion or an implanted medical device. Another concern is the relative vulnerability of the devices. Unlike computers which have the processing power to include antivirus software, patches, etc., many devices within the Internet of Things may include nothing more than a tiny processor chip and a wireless connection with little or no security encryptions built into the chips. These may be easier targets for hackers than standard computer networks. In other words, the cyberwarfare front is widening; and it is still asymmetric. The cost to mount an attack is low relative to both the potential damage it could cause and to the cost of mounting a defence against it.

Government Under Scrutiny

One thing that DHS and FEMA have in common is that they were both created through pulling together several independent government organizations. Indeed, FEMA itself has been a part of DHS since 2003. As the newer of the two organizations DHS is still suffering growing pains, CBP, in particular, is now the keeper of the "single window" of a massive international trade portal. Both DHS and FEMA have faced tough scrutiny by the public, the media, and Congress during the multitude of terrorist incidents and natural disasters that have occurred since each was created. Each has to balance its focus on high disruption and low disruption events and be aware of its responsibilities when events do occur. For example, they have their own version of the disruption timeline as shown in Fig. 8.2. To be effective both DHS and FEMA are reliant on public-private partnerships which need constant nurturing. If a major terrorist event were to occur on US soil involving mass casualties, it would be the litmus test for DHS and FEMA in terms of working effectively together. Since such a test has not occurred as yet, it is a matter of speculation whether or not the law enforcement functions in DHS agencies can coordinate with FEMA's disaster relief planning and efforts. Some even see the two as incompatible and recommend moving FEMA outside of DHS.[1]

The coordination effort which DHS faces is daunting. This would include the sharing of goals, intelligence, and limited funds from the Congress. Under the umbrella of homeland security are the following somewhat disparate functions: emergency management, food safety, immigration, law enforcement (including security and international trade), public health, the president's secret service, and transportation (for passengers and cargo). The coordination needs to occur horizontally across all of these functions within DHS, as well as its counterparts in other countries. It must also occur vertically down through its state counterparts and private partners.

---

[1] See Bullock et al. (2013, p. 618), Brattberg (2012), and Lehrer (2004).

GOALS DESIRED AND GOALS MISSED

Automated Commercial Environment

When and if automated commercial environment (ACE) becomes fully functional it will be one of the most vast and complex communication systems in the world. It will arguably be one of the most vital as well, since it will be used for data mining and decision making over the flow of international trade in goods into the United States. All importers, all carriers transporting the imports, and all intermediaries helping to facilitate this trade will be plugged into a system which will share commercial and security data with 48 partner government agencies (PGAs) as was shown in Table 4.1. While CBP is the gatherer of the information the final decision to release the cargo can involve many of these entities. The discussion of ERP II in Chapter 5 presents a cautionary tale for all users of systems which attempt to map out complex processes. At the same time ACE has the potential to speed the flow of trade since the cargo manifests are submitted electronically before cargo arrival. This interlude is to be used by CBP to perform risk assessment and determine which cargo to speed through and which to hold for more detailed scanning or inspection.

The promise of ACE is to mitigate the trade-off between free flowing trade and supply chain security. It will need to be a careful blending of technology along with human interaction. This means both relationship building and an understanding of risk assessing.

Container Scanning

The goal of Congress to see all inbound containers scanned seems to be elusive. Congressional subcommittees still hold hearings pressing for 100% scanning. DHS, in partnerships with shippers, carriers, and technology providers, has gone on record at these hearings noting that the technology is not available to provide 100% compliance in a reasonable timeframe. According to the Congressional Budget Office the cost to acquire and implement the necessary technology would cost $22–32 billion from 2016–2026.[2] In fact, CBP at the port of Los Angeles maintains only 10 nonintrusive imaging machines suitable for scanning. CBP has 307 machines distributed across several of the 179 largest ports across the United States. Furthermore, many of these machines

---

[2] Hutchins (2016, p. 26).

are aging and in need of replacement.[3] Currently, X-ray scanning takes several minutes per container. In order to avoid a slowdown of trade flows, any new technology desired to scan *all* containers would need to reduce the time involved to seconds instead of minutes. Another decision involves what the new technology should be looking for. If CBP wishes to scan for contraband, illicit drugs, and weapons of mass destruction, the process will take longer and the technology will take longer to develop.

While tracking and sensor-based technologies were discussed in Chapter 5, a combination of the two has been in development in order to make containers "smart." The technology is being used in order to send a signal whenever there is a security breach in transit. In the mid-2000s CBP saw enough promise in the technology to offer a "green lane" benefit to any shipper/carrier which used them in international transportation. However, CBP has since become concerned that the current technology allows for too many false positives. Time will tell if the promise of a truly smart container will be realized.[4]

Technology Reducing Transportation

When an item is in transport it is more vulnerable to theft and to tampering. Today, additive manufacturing (or 3D printing) allows users to "print" new or replacement parts, as well as smaller items, instead of outlaying the time and money to order the part/item from a vendor. Lead times are reduced and dependent only on the speed of the printer.

The 3D printer adds thin layers of plastic or metal-based inks in order to create the finished part/item much in the way a layer cake is made. We should expect the speed, cost, and quality of additive manufacturing to improve in the future. They may be used by both producers and consumers in a fashion similar to how video tape became a substitute for film and expanded from the television studio into millions of living rooms via home video players. It is possible that consumers in the future will buy designs for their printers instead of finished tangible items from hardware stores and other retail outlets. From a supply chain perspective, upstream vendors who use tools and dies, as well

---

[3] Kulisch (2016, p. 27).
[4] For a complete review of smart container technology and changes in policies, see Prokop (2012) and Giermanski (2013, pp. 55–72).

as lathes, would be in less demand. Also, if the lead time to "print" the part/item becomes fast enough, the need to carry inventory would decrease. Currently, however, the cost factor is still high given that industrial capacity 3D printers can be priced in the tens or even hundreds of thousands of dollars. Also, the compatible metals and plastics are more expensive than the traditional material used in tools and dies. Another constraint is that additive manufacturing, by design, is supposed to disperse the manufacturing process and bring it closer to the end customer. Traditional manufacturing benefits from the economies of scale to be had from long production runs. This has served to keep prices of finished goods down; therefore, it will take some time before manufacturing becomes as decentralized as might be supposed.

The future, however, could be one where manufactured items are not moving along domestic and international supply chains. The tangible items that do move along them may shift from high value-added and fragile to bulk plastics, metals, and inks. Many of the traditional targets in theft would be removed. Cargo screening and inspection times may decrease as well when the shipment takes on more of a homogenous and bulk quality. In other words, contraband would stand out even more than today.

Research and testing of autonomous (i.e., self-driving) vehicles are actively being pursued. The variable costs of drivers would be eliminated, as well as the "inside job" potential for theft.[5] Sensors can be used to have vehicles operate closer together and, in effect, move like a virtually connected freight train down a highway. Of course, if the electronic system can be hacked into, the severity of accidents will increase. For individual commercial vehicles there may be the potential to hack into the routing system and send it straight to awaiting criminals.

Remaining Humble Amidst Complexity

When a terrorist event or a natural disaster occurs the first priority is saving or protecting lives in the impacted area. Supply chains at that stage of the process need to be effective in dealing with these challenging and messy situations. Efficiency comes afterward, once

---

[5] Variable costs in for-hire trucking tend to be the largest proportion compared to all other modes of transport. Cost shares in the 70–80% range are not uncommon. For more detail on modal cost structures, see Prentice and Prokop (2016, pp. 103–104).

the situation is stabilized. In this regard it is useful to note that these complex situations need to be planned for but with the humble recognition that black swan events cannot be controlled. They must be reacted to.

With Big Data comes an avalanche of data. But the appropriate mix of data must be determined and adjusted as circumstances warrant. ERP II systems require a mapping of complex activities; and reengineering the process from an "as is" scenario to a "to be" scenario. Both of these systems involve buy-in from a lot of people because these technologies are only as good as the people using them. The role of the human element in the face of technology is crucial.

The lesson is that the decision makers need to be constantly learning, using their imaginations, and thinking outside of the box. The randomness of natural disasters coupled with the need to stay ahead of the imagination of the criminals themselves mean that data, computer firewalls, and other security systems are not a licence for complacency.

Partnership and Policing

The PGAs which handle international trade and the federal, state, and local agencies which handle emergency management all rely on partnership with the private sector. In the trade sector it is a matter of incentivizing regulatory compliance and membership in partnership programs. Emergency management is a matter of having strong vendor relationship to help manage relief and reconstruction efforts.

As discussed in Chapter 6, the Advisory Committee on Commercial Operations (abbreviated COAC) is an important public-private partnership which gives the private sector a voice in how DHS regulates trade and security. Currently, COAC has tilted toward trade facilitation and away from security and policing. Of course, this position could change at any time given COAC's make-up and the state of world trade. The sharing of best practices (e.g., through C-TPAT) should also improve the policy-making process.

A major theme discussed in previous chapters has been balancing partnership and policing. Building trust and cooperation between government and shippers/carriers can lead to Pareto-improvements or socially optimal outcomes but not necessarily stable ones. As seen in Figs. 2.1, 2.2, 8.3, and 8.4, it is challenging to achieve strong partnerships, coordination, etc. However, it is important to open the dialog. This will be an ongoing discussion into the future.

There is a trust and partnership, of a sort, which may emerge between decision makers and their machines (i.e., their Big Data sets and the algorithms computers use in "learning" and predicting). The idea of partnership here is to acknowledge that the process between humans requires trust. Likewise, decision makers should be cautious in relying too heavily on them. Remember, the computers are programmed to find patterns using past data. It is always a leap to go from this to profiling people and situations and forecasting the future. The human touch will always be necessary in these probabilistic processes. In this regard, recall the process discussed in Chapter 5 about how data becomes wisdom. Related to this is the post-incident review as shown in Fig. 8.2. While reassessments and learning are critical after a crime or attack has occurred, they can also be used to assess predictions on a *post facto* basis. Examine the predictions made by the algorithms to see why they were or were not prescient. Was it a matter of data? Was it the protocols in the algorithm? Was it a black swan event? In other words, machine learning will be no substitute for people learning.

GLOBALIZATION

Interconnected

International trade makes it less likely that states will go to war. Businesses which have international connections do not want to see these consumer markets and supply lines cut by military conflict. However, international trade (especially free trade and further economic integration) means that a country does not have complete control of its economic sovereignty. If trade partners gain access it may upset domestic industries which do not want to see the extra competition. Trade agreements, therefore, can often run against the feelings of protectionism which a country's electorate can be prone to in times of economic slowdown. Politicians may find themselves in the tenuous situation of trying to represent the will of the people while, at the same time, trying to hold them to preexisting trade agreements. Therefore, international trade, sovereignty, and democracy all face a tension when coexisting.[6]

---

[6] For an examination of this trilemma, see Rodrik (2011).

In fact, it is not possible to increase the magnitude of all three. Any two of these activities will come at the expense of the remaining. The right balance is a political decision. Within this process is the government's policy regarding supply chain security. If international trade is subject to tension, then, so are all security policies surrounding it.

Supply chain security can be used for contradictory purposes. Those who wish to see international trade and economic integration proceed may concede, in an age of crime and terrorism, that a measure of supply chain security programs is necessary to assure the public of a safe flow of trade. On the other hand, supply chain security programs can be used to advance protectionist policies. For example, the United States' flat ban on Mexican trucks proceeding inland beyond 25 miles because of supposed safety concerns is a convenient mask for minimizing Mexican competition on the roadways. Of course, Mexico reciprocated with a similar ban on US trucks in Mexico. Another example is the requirement for only US-flagged vessels to engage in cabotage along the coast and river ways. This "Jones Act" requirement has been in place for decades and its rationale is that in time of national emergency when nearby vessels are commandeered they will be US-flagged and, therefore, part of a US company. In the meantime, however, the US coast and riverways are not subject to foreign competition for point-to-point domestic transportation.[7]

The trilemma of trade, sovereignty, and democracy means that there is no stable outcome to be expected as geopolitical debates and electoral cycles shape the economy. Also, it is too simple to think about trade in terms of final goods being traded between two countries and with each good being produced in full in the respective country. Raw material, subassemblies, and final goods can be sent among various countries due to global supply chain realities.

Global Trade Versus Regional Trade Blocs

The world is caught between two trends: a globalized trading system and regional trade. The World Trade Organization (WTO) exemplifies a collection of countries wishing to let a global organization monitor trade flows and arbitrate disputes. The WTO is the business and trade equivalent of the United Nations. On the other hand, the world is also splintered into regional trade blocs, which seek free trade among their

---

[7] For a complete review of cabotage regulations in the United States, see Prokop (2014).

members but not necessarily among the various blocs. Prominent examples are the North American Free Trade Area (NAFTA) and the European Union (EU).

However, the supposed irresistible path of globalization is starting to see resistance. From Britain's decision to exit the EU to the election of Donald Trump as president, on an "America First" platform, many voters are worried about an unchecked process of globalization. Substituting free trade for protectionism or for some notion of fairer trade will generate more uncertainty in international business affairs as governments sort out the will of their people.[8]

A Global or Fractured Internet

Just as physical trade flows can be slowed by protectionism, so can the Internet. For years countries such as China, Russia, and Iran have set up firewalls to block certain content. While cyberspace is hard for one country to regulate, it is easier to target the brick-and-mortar operations of Internet-based companies located in the various countries as a way to pressure for cyber-regulations. A free-wheeling Internet has been a virtual superstructure for information sharing at low cost. Of course, this has come at a price. Public and governments fear an invasion of privacy, espionage, cybercrime, and cyberwarfare. Pressure is building in some jurisdictions for more Internet regulation. It can come in two forms: (1) regulating its structure such as addresses and technical standards[9]; and (2) regulating who uses it and how. Point (2) is where many governments and courts get involved. Political causes such as freedom of expression and human rights can be fought over through the lobbying or litigation of special interest groups. Another prominent example is the enforcement of "right to be forgotten" laws (e.g., upheld by the European Court of Justice in 2014). All of these trends will shape the

---

[8] Of course, free trade does not mean *free* trade. Even free trade partners can get into disputes over import dumping, immigration, and foreign direct investment. Furthermore, free trade agreements tend to involve the lowering of trade tariffs and leave untouched many non-tariff trade barriers. In fact, compliance with supply chain security regulations is itself a non-tariff trade barrier. These are, by design, supposed to add to the transaction cost of trade.

[9] Currently, the Internet Corporation for Assigned Names and Numbers (ICANN) oversees web addresses and the Internet Engineering Task Force (IETF) oversees technical standards. Both of these are nonprofit organizations.

way information is traded online and stored in the Internet Cloud. Just as countries have trade barriers applicable to trade in goods, services, capital, and people, so might such transaction costs apply to trade in information.

CONCLUSIONS

Supply chains are systems and securing them properly means to secure whole systems as opposed to individual nodes in isolation. Since any physical or cyber-based attack or theft will ripple both upstream and downstream along the supply chain, any vulnerability with one organization leads to vulnerability among the partners. In this context the way to think about supply chain security is in terms of how the system works, where it is more probable to suffer attack or theft, and how quickly it can recover. This book has shown the importance of government as a player along the supply chain, as well as the role of international trade in supply chain management. The government can be a partner or a police officer. International trade brings in multiple other governments into the supply chain on top of the shippers, carriers, and intermediaries, which may be based in other countries. Global supply chain security is particularly challenging. It may well be the most important public-private partnership for business and society.

Bibliography

Brattberg, E., 2012. Coordinating for contingencies: taking stock of post-9/11 homeland security reforms. Journal of Contingencies and Crisis Management 20 (2), 77–89.

Bullock, J.A., Haddow, G.D., Coppola, D.P., 2013. Introduction to Homeland Security, fourth ed. Butterworth-Heinemann, Waltham, MA.

Giermanski, J.R., 2013. Global Supply Chain Security. Scarecrow Press, Inc., Plymouth, UK.

Hutchins, R., July 25, 2016. 100 percent fired up: the debate over scanning of all U.S.-Bound containers flares up again in Congress. Journal of Commerce 26.

Kulisch, E., September 2016. The 100% option. American Shipper 22–27.

Lehrer, E., 2004. The homeland security bureaucracy. Public Interest 156, 71–85.

Prentice, B.E., Prokop, D., 2016. Concepts of Transportation Economics. World Scientific Publishing, Singapore.

Prokop, D., 2012. Smart containers and the public goods approach to supply chain security. International Journal of Shipping and Transport Logistics 4 (2), 124–136 Special Issue on Container Security and Supply Chain Visibility.

Prokop, D., 2014. International transportation management. In: Prokop, D. (Ed.), The Business of Transportation, vol. 2. Praeger, Santa Barbara, CA, pp. 70–89. Applications.

Rodrik, D., 2011. The Globalization Paradox: Why Global Markets, States, and Democracy Cannot Coexist. Oxford University Press, Oxford, UK.

Appendix: Some Useful Concepts and Analytical Tools

OUTLINE

This appendix covers some concepts and analytical tools, along with suggested readings, for those who need a review of undergraduate transportation management and industrial organization. Key terms used in contracting are discussed in order to establish lines of responsibility. The five modes of transport are outlined in terms of their cost structure. Efficiency and effectiveness are compared and contrasted. Finally, the meaning of free trade and the extent to which other barriers remain are discussed. All of these terms and concepts are examined in the context of supply chain security in order to clarify points made in previous chapters.

INTERNATIONAL COMMERCIAL TERMS AND THE UNIFORM COMMERCIAL CODE

International Commercial Terms (INCOTERMS) were developed by the International Chamber of Commerce to help facilitate international trade. The Uniform Commercial Code (UCC) is a United States construct to facilitate interstate commerce. The UCC is overseen by the National Conference of Commissioners for Uniform State Laws. It is important to note that neither of these systems constitutes law and, as such, using their terms is optional in contracts. However, each system collects established legal terms and conveniently categorizes them. Of course, the advantage in using them is that they sharpen the meaning of a contract since the terms have a long history of use in commercial transactions. In contracting it is always a good idea to avoid ambiguities or lengthy verbiage in place of well-understood commercial terms.

Table A.1 UCC Articles

Title	Contents
Article 1—General provisions	Definitions, rules of interpretation
Article 2—Sales	Sales of goods
Article 2A—Leases	Leases of goods
Article 3—Negotiable instruments	Promissory notes and drafts (commercial paper)
Article 4—Bank deposits and collections	Banks and banking, check collection process
Article 4A—Funds transfer	Transfers of money between banks
Article 5—Letters of credit	Transactions involving letters of credit
Article 6—Bulk sales	Auctions and liquidations of assets
Article 7—Documents of title	Storage and bailment of goods
Article 8—Investment securities	Securities and financial assets
Article 9—Secured transactions	Transactions secured by security interests

When reading through these variations, it is worth keeping in mind the different ways buyers, sellers, and the carriers of the sellers' goods interact. They represent a small part of the process mapping discussed in Chapter 5, and they show when ownership or responsibility is transferred among the parties. Those taking ownership or bearing responsibility also bear the onus of taking on security as well.

The UCC is composed of nine articles which cover many activities which are documented in a contract (see Table A.1).

Eleven terms constitute the latest version of INCOTERMS.[1] These are divided into those which cover all modes of transport and those applicable to maritime (ocean and inland waterway).

TERMS FOR ANY TRANSPORT MODE

The first two terms require the buyer of the goods to contract and pay for the transport.

- EXW—EX WORKS (… named place of delivery)
  - The seller's only responsibility is to make the goods available at the seller's own premises. The buyer bears the full costs and risks of moving the goods from there to the buyer's destination.
- FCA—FREE CARRIER (… named place of delivery)

---

[1] For a complete discussion see Ramberg (2011).

- The seller delivers the goods, cleared for export, to a carrier selected by the buyer. The seller loads the goods only if the carrier pickup is at the seller's premises. From that point, the buyer bears the costs and risks of moving the goods to the buyer's destination.

The next two terms require the seller of the goods to contract and pay for the transport.

- CPT—CARRIAGE PAID TO (… named place of destination)
  - The seller pays for moving the goods to the buyer's destination. From the time the goods are transferred to the first carrier, the buyer bears the risks of loss or damage.
- CIP—CARRIAGE AND INSURANCE PAID TO (… named place of destination)
  - The seller pays for moving the goods to destination. From the time the goods are transferred to the first carrier, the buyer bears the risks of loss or damage. The seller, however, contracts and pays for the cargo insurance.

The next three terms relate to how the goods are delivered.

- DAT—DELIVERED AT TERMINAL (… named terminal at port or place of destination)
  - The seller accomplishes delivery when the goods, once unloaded from the arriving means of transport, are placed at the buyer's disposal at a named terminal at the named port or place of destination. "Terminal" includes any place, whether covered or not, such as a quay, warehouse, container yard or road, rail or air cargo terminal. The seller bears all risks involved in bringing the goods to, and unloading them at, the terminal at the named port or place of destination.
- DAP—DELIVERED AT PLACE (… named place of destination)
  - The seller accomplishes delivery when the goods are placed (though not unloaded) at the buyer's disposal on the arriving means of transport and is ready for unloading at the named place of destination. The seller bears all risks involved in bringing the goods to the named place.
- DDP—DELIVERED DUTY PAID (… named place)
  - The seller accomplishes delivery when the goods are cleared for import and ready to proceed to the buyer's named place of destination. The seller bears all costs and risks of moving the goods to destination, including the payment of any customs duties and taxes.

TERMS FOR MARITIME (OCEAN AND INLAND)

The first two terms require the buyer of the goods to contract and pay for the transport.

- FAS—FREE ALONGSIDE SHIP (… named port of shipment)
  - The seller delivers the goods to the origin port alongside the ship nominated by the buyer. From that point, the buyer bears all costs and risks of loss or damage.
- FOB—FREE ON BOARD (… named port of shipment)
  - The seller delivers the goods to the origin port and arranges to load them on board the ship nominated by the buyer. The seller also clears the goods for export. From that point, the buyer bears all costs and risks of loss or damage.[2]

The last two terms require the seller of the goods to contract and pay for the transport.

- CFR—COST AND FREIGHT (… named port of destination)
  - The seller clears the goods for export and pays the costs of moving the goods to the buyer's port of destination. The buyer bears all risks of loss or damage.
- CIF—COST, INSURANCE, AND FREIGHT (… named port of destination)
  - The seller clears the goods for export and pays the costs of moving the goods to the port of destination. The buyer bears all risks of loss or damage. The seller, however, contracts and pays for the cargo insurance.

TRANSPORTATION CARRIERS

The five major modes of transportation are pipelines, railroads, ocean vessels, airlines, and motor carriers. While they are not homogenous, their differences are not too hard to comprehend. Their biggest differences are in terms of cost structure and level of competition. Simply put, if fixed costs are a large proportion of the total costs faced by the carrier these act as a barrier to entry which serves to reduce competition. Fixed

---

[2] Technically, the INCOTERM would be written as "FOB port of origin." It is important to note that FOB is not a stand-alone term. It needs to have a place marker with it since that is where the freight price is set and title of ownership is passed. For a full discussion on the use of the FOB term see Prokop (2014a).

costs in transportation are best seen in terms of the infrastructure the carrier has to pay for up front in order to start business and maintain going forward in order to stay in business.

A pipeline company has to build its entire operation. These include storage facilities, pumping stations, and the pipeline itself. A railroad has to build tracks, switches, yards, and terminals before its locomotives and railcars can begin operations. Both of these modes build their own right-of-way networks. Ocean vessel companies and airlines have to spend a lot of money for their conveyances; however, they do not have to build their own ports of call. Sea ports and airports are typically built by government entities. The ocean vessel company may rent space at the sea port and own and operate heavy equipment like gantry cranes. Likewise, an airline may rent space at airports in order to set up hangars and book landing slots in order to service passengers and/or air cargo. These two modes have large proportions of fixed cost but nowhere near the levels found in pipeline and rail operations. Rights of way are handled by the sea ports and airports themselves. The mode with the lowest proportion of fixed cost is the motor carrier or for-hire trucking mode. To start a business, all that is needed is a truck and a licensed driver. The motor carrier may require terminals and yards but for the most part the infrastructure and right-of-way are provided in the form of public roads, tunnels, and bridges.

From a security perspective more infrastructure means more areas of vulnerability and, therefore, more details necessary in security planning. However, proceeding from pipelines to motor carriers shows an increasing level of shared responsibility in security planning. This may be a good thing or a bad thing depending on the level of trust and cooperation among the partners. Partnerships are useful when they expand the range of experiences and collective imagination brought to the planning process. However, they are not useful when they slow down decision making and muddle any attempt to consensus-building. Also, government regulates each mode of transport differently and to different overall degrees. For example, choices of routes in international trucking are not regulated while they are in international airlines.[3] This means the level of partnership with the government in security planning will necessarily differ.

---

[3] For a review of the unique structure of international airline regulation, see Prokop (2014b).

EFFICIENCY VERSUS EFFECTIVENESS

Efficiency has a precise meaning in economics. It has two components: productive efficiency and allocative efficiency. An organization needs to have both components satisfied in order to be efficient in the proper sense of the term. Productive efficiency means to perform a task in a least-cost manner. This means there is no other method or combination of inputs to produce the output without adding to the cost of production. In this way there is no waste. But an important companion is the task of allocating resources to the production of an item that is desired in the marketplace for the price asked. Thus, allocative efficiency means that the item produced is both desired and represents the best use of the organization's limited resources. In other words, the opportunity cost is acceptable.

From a security perspective, suppose a business acquires a security network which provides real-time tracking and tracing of all items in transit. As a result thousands of data points are generated every minute, nonstop. If the business was able to secure a very low price for this technology because of a good relationship with the vendor a case for productive efficiency in data generation can be supported. As to allocative efficiency it depends on how the technology is to be used downstream along the supply chain. Can the data indicate a security breach when it happens? Are protocols in place to deal with a breach? Are the customers willing to pay extra charges for the added benefit of this service (since the business needs to recoup its costs)? If the answer is no to one or more of these questions then the system is not efficient. It would represent a misallocation of resources. Another way to look at this is to install a security system that is fool-proof but so expensive that the business's product is priced out of the market. In summary, efficiency is a balancing act between cost control and satisfying market demand in terms of both quality and price.

Effectiveness is not the same thing as efficiency. Effectiveness means that an organization is being good enough under the circumstances. From a security perspective it means that the affected organizations along the supply chain are managing any disruptions. They are dealing with the mess and chaos which ensue, all for the purpose of trying to stabilize the situation. Only when stability is reestablished can normal market forces resume. Once a market is functioning properly, a return to efficiency can be contemplated by the organizations.

FREE TRADE AND NON-TARIFF TRADE BARRIERS

Free trade does not really mean "free trade." Free trade agreements typically involve the lowering, or complete removal, of tariff barriers on importing of inputs and final goods. These tariffs act as sales taxes on imports and removing them does make them more competitive. However, paperwork remains even for importers and exporters in free trade partner countries. Often free trade agreements maintain elaborate "rules of origin" which dictate how much production of an item must take place in a partner country if it is to be allowed to enter duty-free. Today, a lot of manufactured items are produced in stages across multiple countries many of which are not free trade partners of the United States.

Many non-tariff barriers remain in place under free trade agreements. These serve to keep trade anything but free. Free trade partners may still insist on domestic content requirements, banning of certain chemicals/ingredients, and subsidizing certain industries (domestic agriculture being a popular one within North America and Europe). Also, while free trade agreements tend to focus on inputs and final goods, they seldom ease the process of trade in other areas (e.g., workers and financial capital).

Compliance with supply chain security programs is a non-tariff trade barrier to trade as far as those shippers, carriers, and intermediaries wishing to ship a product into the United States are concerned. It is important to note that non-tariff barriers slow trade but that does not make them inherently bad. However, they need to be justified based on some political or economic purpose. Chapter 4 noted supply chain security programs that did require time and money in order to be in compliance. However, Chapter 7 noted that programs, if designed well, might lead to a positive-sum outcome. If security programs can increase the flow and certainty of trade then, as non-tariff barriers, they can serve an economic purpose.

Bibliography

Prokop, D., 2014a. Transportation in business decision-making. In: Prokop, D. (Ed.), The Business of Transportation. Modes and Markets, vol. 1. Praeger, Santa Barbara, CA, pp. 1–14.

Prokop, D., 2014b. Government regulation of international air transportation. In: Peoples, J. (Ed.), The Economics of International Airline Transport. Advances in Airline Economics, vol. 4. Emerald Group Publishing Limited, Bingley, UK, pp. 45–59.

Ramberg, J., 2011. ICC Guide to Incoterms 2010. ICC Services Publications, Paris, France.

Index

'*Note:* Page numbers followed by "f" indicate figures and "t" indicate tables.'

189